Symbols of excellence

By the same author from Cambridge University Press:

Excavations at Star Carr
World Prehistory in New Perspective
The Earlier Stone Age Settlement of Scandinavia

The Imperial State Crown

Symbols of excellence

Precious materials
as expressions of status

Grahame Clark

Emeritus Disney Professor of Archaeology,
University of Cambridge

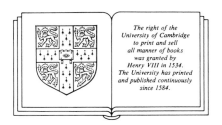

The right of the
University of Cambridge
to print and sell
all manner of books
was granted by
Henry VIII in 1534.
The University has printed
and published continuously
since 1584.

Cambridge University Press

Cambridge

London   New York   New Rochelle
Melbourne   Sydney

Published by the Press Syndicate of the University of Cambridge
The Pitt Building, Trumpington Street, Cambridge CB2 1RP
32 East 57th Street, New York, NY 10022, USA
10 Stamford Road, Oakleigh, Melbourne 3166, Australia

First published 1986

Printed in Great Britain by
BAS Printers Limited, Over Wallop, Stockbridge, Hants.

British Library cataloguing in publication data
Clark, Grahame
Symbols of excellence: precious materials as
expressions of status.
1. Gems 2. Precious metals 3. Symbolism
I. Title
553'.8   GR805

Library of Congress cataloguing in publication data
Clark, Grahame, 1907–
Symbols of excellence.
Bibliography: p.
Includes index.
1. Gems. 2. Precious metals. 3. Symbolism.
4. Social status. I. Title.
GT2250.C55 1985   394 85–16591

ISBN 0 521 30264 1

Contents

List of illustrations   vii

Preface   ix

1 *Introductory*   1

2 *Organic materials*   13

3 *Jade*   33

4 *Precious metals*   50

5 *Precious stones*   65

6 *The symbolic roles of precious substances*   82

Notes on the colour plates   107

Notes   110

Index   120

Illustrations

Colour plates   *facing page*

Frontispiece The Imperial State Crown   iii
A Amber objects from Mesolithic Denmark   4
B Maori adze with nephrite blade   5
C Egyptian Middle Kingdom necklace   5
D Sumerian court jewellery   22
E Mosaic mask of Quetzalcoatl   23
F Maori *tiki* of greenstone   38
G Romanesque crozier of walrus ivory   39
H Tutankhamun's innermost coffin   46
I Pectoral from the tomb of Tutankhamun   47
J The 'Phoenix' portrait of Queen Elizabeth I   62
K The Chancellor of the University of Cambridge   63

Figures   *page*

1 Upper Palaeolithic burial, Sunghir, USSR   2
2 The use of precious substances through time   5
3 The occurrence of Aegean *Spondylus* shells in Neolithic central Europe   8
4 Ivory head from Brassempouy, France   16
5 The Barberini Ivory   18
6 Minoan snake goddess   20
7 Rhinoceros horn cup, Ming dynasty   21
8 Skull set with cowries from Jericho   24
9 Sepik river skull with cowrie eyes   25
10 Jet necklace of the Scottish early Bronze Age   32
11 The main sources of jade used in China   36
12 Adze blade carved from nephrite boulder   38
13 Chinese jade *zong*   40
14 Hardstone *bi*-ring   41
15 Hawk pendant from An-yang   43

16 Jade burial garment of Dou Wan   44
17 Olmec votive groupfrom La Venta   48
18 Mycenaean gold cup from Vapheio   55
19 Gold torc from Snettisham   56
20 Gold bracteate from Ravlunda   57
21 Gold collar from Alleberg   58
22 Roman silver cup from Hoby   61
23 Tang dynasty silver cup   62
24 Mycenaean dagger from Gournia   63
25 The spread of lapis lazuli from Afghanistan   66
26 Classical cameo of Augustus   71
27 Detail of the Sutton Hoo purse lid   74
28 Rose and brilliant diamond cuts   76
29 'Founder's Jewel', New College, Oxford   79
30 Silver cauldron from Gundestrup   86
31 Reliquary statue of Ste Foy, Conques   89
32 Jewelled cover of the 'Book of the Pericopes'   90
33 Chalice of Abbot Suger of Saint Denis   92
34 Archbishop Maximian's throne, Ravenna   94
35 The Holy Crown of Hungary   95
36 Richard, Earl of Cornwall's Crown   96
37 Catherine the Great's mitre crown   98
38 The Sceptre with the Cross   99
39 (a), (b) Greek coin of Ephesus   103
     (c), (d) Roman *denarius*   103
     (e), (f) Roman *aureus*   103
40 Plate of the London Goldsmiths' Company   104
41 The Derby Trophy   105
42 The Football Association Cup   105
43 Olympic medal, 1908   106

Preface

It is a pleasure to acknowledge the help afforded by my publishers not merely in reducing errors, but more positively in offering suggestions to make this book more enjoyable to potential readers. I am particularly indebted to my editor, Dr Peter Richards, for his interest and creative suggestions, to my sub-editor Dr Caroline Murray who has spared me many infelicities and to Mr Dale Tomlinson for his part in designing the book and helping to ensure the most effective conjunction of illustrations and text. I would like to thank the librarian of New College, Oxford, Dr G. V. Bennett, for putting me right over the historical antecedents of the so-called Founder's Jewel of New College; and Mr Albert Batteson of Ede and Ravenscroft Ltd for information about the robes illustrated in Plate K. To my wife I am deeply grateful for going over the text, spotting errors which had escaped previous scrutiny and raising questions which drove me back to my sources.

Finally I would like to make formal acknowledgement to the following for the use of photographs: All Sport/Photographic Ltd, Figs. 41, 42; Ann Münchow, Aachen, jacket illustration, Fig. 36; Bildarchiv Foto Marburg, Fig. 35; British School of Archaeology in Jerusalem, Fig. 8; Cambridge Evening News, Plate K; Crown Copyright, Frontispiece, Fig. 38; Hirmer Fotoarchiv, Figs. 5, 18, 22, 32, 34; Joubert/SPADEM, Fig. 31; London Goldsmiths' Company, Fig. 40; Lunds Universiteits Historiska Museum (courtesy Martha Strömberg), Fig. 20; Musée des Antiquités Nationales, St Germain-en-Laye, Fig. 4; Museum of Fine Arts, Boston, Fig. 6; National Gallery of Art, Washington D.C., Fig. 33; Nationalmuseet, Copenhagen, Plate A, Fig. 30; National Museum of Antiquities of Scotland, Edinburgh, Fig. 10; National Portrait Gallery, London, Plate J; New College, Oxford, Fig. 29; Novosti Press Agency, Figs. 1, 37; Östasiatiska Museet, Stockholm, Fig. 7; Robert Harding Picture Library, Plates H and I; Statens Historiska Museum, Stockholm, Fig. 21, top and bottom; Trustees of the British Museum, Plates B, C, D, E, G, Figs. 13, 14, 19, 26, 27, 39a–f, 43; University Museum of Archaeology and Anthropology, Cambridge, Plate F, Fig. 24; Xinhua News Agency, Fig. 16.

Grahame Clark

1

✧✧✧

Introductory

As I have recently sought to emphasise, we owe our identity as human beings to the fact that by contrast with other animals we belong to societies constituted by sharing values.[1] It is because values can be most effectively defined and expressed in words that the study of literature claims a central place in humane studies. Yet literature can only give access to the values entertained by the members of literate communities and in these only for persons able to apprehend what they read. Where written records do survive they are as a rule woefully incomplete, and for prehistoric communities they are by definition completely absent. It was precisely during these remote and largely unrecorded periods that some of the most crucial changes took place. To see how men began to entertain values and engage in the pursuit of excellence we need additional sources of information. Fortunately archaeology is capable of providing these in abundance. Ever since men, whether they could read or write, have entertained values they have sought to express these in symbols. Accepting the definition of a symbol given in the *Shorter Oxford Dictionary* as 'a material object representing or taken to represent something immaterial or abstract', one may safely look to archaeology, a discipline whose very existence depends on its ability to learn from things what is not available in words.

Prehistoric archaeologists the world over have increasingly focussed attention on the ecological and economic aspects of life in antiquity. Intensive research has been directed to the ways in which early communities adapted to and utilized their environments. Above all, attention has concentrated on the ways in which in the course of time they managed to transform their basis of subsistence, increase and concentrate populations and by making possible a finer subdivision of labour promote advances in technology which in turn generated further cycles of progressive change. Fruitful though this has been in promoting and giving direction to research, its uncritical pursuit only serves to advance a reductionist and therefore an inadequate view of man. The communities which have left archaeological deposits are necessarily ones which have survived in competition with others, yet their capacity to survive depended not merely on the effectiveness of their economies and technology, but also on maintaining standards of

1

Figure 1 Upper Palaeolithic burial, Sunghir, USSR. Already in Upper Palaeolithic times the dead were buried with their personal finery. The illustration shows the head and trunk of a young boy wearing a number of bracelets and many beads, most of which were probably attached to clothing.

excellence. It is worth emphasizing that, as Colin Renfrew noted in the context of early metallurgy in the Balkans,[2] technology itself has not invariably been developed to serve material ends. The finest bronze castings of Shang China and arguably of all time were not made for implements or even weapons, but for the ritual vessels centred on the cult of ancestors.[3] Again, the astonishingly sophisticated metallurgy developed in the Andean region during the two millennia before the Spanish Conquest was not motivated to satisfy material appetites. Instead it was developed to promote value systems in the spheres of religion and politics.[4] The early smiths were primarily concerned with the colours and surface appearance of gold and silver. In order to gain the maximum effect they manufactured alloys of copper, silver and gold and invented procedures for gilding and silvering copper. The sophisticated systems of electro-chemical replacement and depletion gilding and silvering developed by Andean metallurgists during the first millennium before Christ were designed first and foremost as symbols of rank, power and religious force. It is because human societies pursue values that their study falls within the province of history rather than biology.

The ability to discriminate is admittedly a basic ingredient of survival. Choosing the most appropriate things to eat and in the case of man the most effective materials for tools and weapons is adaptive in the sense that it makes for biological survival: organisms which practise the keenest discrimination flourish at the expense of those less discriminating in their choices. In the same way men have enhanced their success as human beings rather than as mere primates by exercising discrimination in their choice of symbols of excellence. Men are the only animals to set store by discovering, acquiring and displaying materials comparatively rare in nature, frequently only to be obtained from distant sources and commonly useless for the purposes of daily life. By designating such materials as in varying degrees precious they have created symbols of excellence, a quality which stems from aesthetic awareness but the striving for which lies at the very root of the civilizations created by man.

The finding of archaeology that over the last five thousand years men of the most diverse civilizations have invariably set the highest values on substances which, however attractive aesthetically, were nevertheless useless for purposes of daily life, coincides with the observation of North American society during the last quarter of the nineteenth century made by Thorstein Veblen and embodied in his classic book *The Theory of the Leisure Class*, originally published in 1899.[5] Although composed with the animus and spleen of a man condemned by his personality to a life of persistent failure, Veblen produced a book hilarious in style but of quite brilliant perversity. Although still widely read, more especially in the United States, as a satire on 'the emptiness of capitalist philosophy' (to quote from the blurb on the binding of a recent edition), the book in fact offers an insight into the development of civilized societies of particular relevance to the social orientation of modern archaeology. By insisting[6] that 'with the exception of the instinct of self-preservation, the propensity for emulation is probably the strongest and most alert and persistent of the economic motives proper' he in effect undermined one of the basic assumptions of ensuing generations of archaeologists and more especially of prehistorians, namely that cultural advance can be adequately accounted for primarily in terms of utility at the level of subsistence and technology.

Veblen's central thesis was that emulation took the form of engaging in a conspicuous way in honorific as opposed to merely useful pursuits. Honour was to be obtained by engaging in activities and indulging interests which were essentially and palpably useless. In summary terms emulation proceeded by indulging directly or vicariously in conspicuous waste. It is particularly relevant to the present work, concerned with the role of precious substances as symbols of excellence, that is as objects of emulation, that Veblen should have specified 'precious stones and the metals and some other materials used for adornment and decoration' in so many words as being of outstanding 'utility as items of conspicuous waste'. Furthermore, it is interesting that he held that they owed this 'to an antecedent utility as objects of beauty'.[7]

Veblen took pleasure in representing that in a society as dedicated to practical efficiency as the United States emulation should have been directed to the pursuit of objectives as conspicuously useless and therefore wasteful as gold or precious stones. His wrong-headedness resided in his failure to recognize that the attitudes and expenditures of which he made so much fun were merely symbols of achievement, the kind of achievement which has in fact given rise to every civilization and marked stages in the development of each one. Again, he failed to appreciate that the conspicuous consumption which was the fruit of successful emulation was not as invariably absurd as he made out, or that in the course of history world-wide they were responsible for the greatest architecture and other manifestations of high art.

As a recent and fundamentally sympathetic editor, Professor C. Wright Mills, put it, Veblen was blinded by the assumption quoted from his own book that 'the accumulation of wealth at the upper end of the pecuniary scale implies privation at the lower end of the scale'.[8] In reality, according to Wright Mills, Veblen did not write as he imagined a theory of general application, but one directed to criticism of the *nouveau riche* of North America during the closing decades of the nineteenth century. Furthermore he wrote from the standpoint of one who had failed conspicuously to emulate his own colleagues during his lifetime. Even so

A Amber objects from Mesolithic Denmark

B Maori adze with nephrite blade

C Egyptian Middle Kingdom necklace

it is strange that Veblen should have failed to recognise the magnitude of what the fierce emulation of railroad promoters, industrialists and bankers in fact managed to achieve in North America during so short a space of time. It was rivalry that tamed a continent, harnessing the resources of harsh environments and creating an economy that before long was to give the United States the industrial leadership of the world. If Veblen's book continues to be read this is partly for its unconscious humour but partly for the comfort, however misplaced, which it offers to those still mesmerized by egalitarian ideas.

A ready way to identify substances which in the course of many thousands of years have come to be accepted as precious is to scan the windows of jewellers' shops in Bond Street or the Burlington Arcade and their counterparts in the wealthiest cities of the western world. This continuity is hardly to be wondered at when it is remembered that precious substances have attracted and continue to attract interest by reason of their inherent qualities. However much the rela-

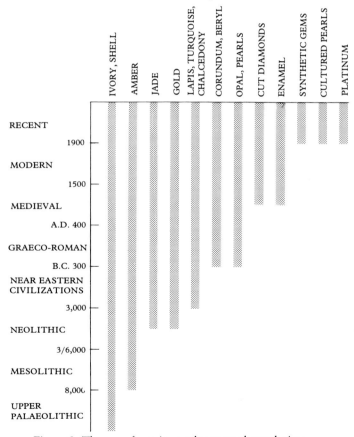

Figure 2 The use of precious substances through time.

tive values of particular substances vary from one culture to another, each one owes its status to physical attributes. Ivory attracts by reason of its smoothness and creamy colour, jade by its translucence, hues and touch, gold by its untarnishable gleam, pearls by their orient and the most precious stones by their fire. It is engaging to reflect when, where and for what reason the various substances featured in modern displays acquired their status as precious substances. One reason why it is worth asking such questions is that ready answers may be found. Precious objects are particularly visible in the archaeological record. Durability has at all times been a main attribute of the materials held in highest regard and it is precisely the most durable things that have most chance of surviving. They also have the best chance of being found. Archaeologists are as keen to discover precious things, though for different reasons, as treasure-seekers. Again, because treasures served to denote prominent persons, they are most likely to be listed or even described in written records, as well as featuring in works of art. Furthermore many fine pieces remain in use today as regalia or ecclesiastical accessories. The history of precious substances is richly documented, principally by archaeology, as far back as their first recognition by some of the earliest representatives of the modern species of *Homo sapiens*.

The concept of precious as distinct from merely useful substances could only have arisen in societies enriched by aesthetic sensibilities and sufficiently aware of persons to wish to symbolize relations between them as individuals and as enactors of social roles. The first communities combining these prerequisites were those which emerged towards the end of the Pleistocene. Aesthetic sensibility was almost certainly older than its earliest expression in symbolic art. It has often been observed that the flint and stone industries of the Middle and early Late Pleistocene, the Lower and Middle Palaeolithic assemblages of prehistory, developed in the direction of greater comeliness as well as greater technical efficiency. Indeed it was the sheer beauty of some of the finer implements from these remote periods which helped to excite the interest of the collectors who laid the foundations for the scientific study of prehistory. There should be nothing surprising about this. Well designed artefacts are not merely more attractive to look at. They are also likely to be more efficient. Those who advocate a greater emphasis on technology and depreciate the importance of art in education might remember that even the mathematical formulae used by engineers are habitually judged in terms of their elegance. The very process of humanization can now be viewed as a product of aesthetic feeling as well as of mind. Michelangelo was no mere wonder. He was an exemplar.

The first undoubted works of art documented by archaeology, the engravings and paintings on the walls and ceilings of caves and rock-shelters together with the figurines and engraved pieces from deposits resting on their floors, were created by members of our own species *Homo sapiens sapiens* during the Late Glacial period from around thirty thousand years ago. The life-like character of much of this art is evident, but few if any authorities would interpret it today as representation for its own sake. André Leroi-Gourhan has recently written:[9]

The parietal assemblages have the essential characteristics of a message: they respond to the needs and means that man has had since the Upper Palaeolithic to produce oral symbols in a material form by using his hands.

Among the many reasons for treating palaeolithic cave art as symbolic is the frequency with which representations are superimposed on the same rock surface, a practice used by the Abbé Breuil to construct sequences in the development of the art. Another, recently insisted upon by Professor Leroi-Gourhan, is the way representations of different sexes and species are disposed within the passages and chambers of the palaeolithic caves. Again, there is the fact that a variety of signs which, whether or not they suggest weapons, huts or traps or remain enigmatic, can only be interpreted as symbolic in intent. Another and in some respects even more powerful exemplification of the use of symbols by Upper Palaeolithic and Mesolithic man is to be found in hitherto little noticed marks on antler and bone artefacts recently subjected to intensive study by Alexander Marshack[10] with the aid of high-magnification photography.

Whether, as many prehistorians suspect, the Upper Palaeolithic was a time of notable progress in articulate speech is something which archaeology can hardly determine. What is certain is that this was a time of decisive advance in the recognition accorded to individual persons. This can be well illustrated from burials. Many burials of Neanderthal man have been carefully excavated in Europe and Palestine, but not a single one was accompanied by personal ornaments. By contrast numerous Upper Palaeolithic interments extending from western Europe to Siberia were sufficiently respected to be buried clothed and accompanied by their personal finery. This included the basic forms of jewellery – necklaces, pendants, bracelets and pins.

If the recognition of persons was a precondition for jewellery, it is no less true that the prime purpose of wearing it was to symbolize status. It follows that the nature of jewellery is bound to reflect in some measure the structure and activities of the society in which it was current. In other words the history of jewellery and of the precious substances incorporated in it needs to be studied in the context of social history and vice versa. The first communities to use it were still small, organized on a segmental basis and dependent on a lithic technology. At this stage the prime role of precious substances was to designate ethnic identity and grade individuals in terms of age and sex. Again, methods for displaying precious substances were of an elementary character.

Among the Upper Palaeolithic and Mesolithic hunter-foragers of Europe and contiguous parts of Africa and Asia precious substances were invariably of organic origin. Much the most prominent in the archaeological record were ivory and shells, and it was mainly from these that the bracelets, necklaces and pendants buried with the dead were made. The adoption of an economy based on farming did not effect a clean break. The ornaments recovered from Neolithic cemeteries in the Rhine and Danube basins show that the shells most favoured for personal adornment were those of the mussel *Spondylus gaederopus* from which were fashioned necklaces, pendants and bracelets. Recent research has confirmed that the shells used in central Europe came in fact from the Aegean.[11]

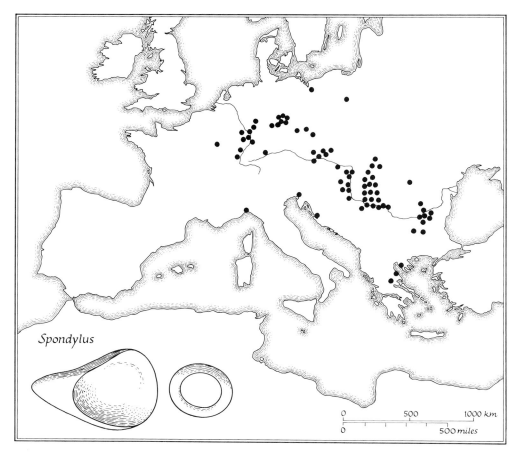

Figure 3 The occurrence of the Aegean mussel *Spondylus gaederopus* in Neolithic central Europe.

From this source they spread by exchange networks as far as north Poland and the Ile-de-France over a radius as the crow flies of more than 1700 km. If these early farmers were able to satisfy their biological needs from resources obtained within an hour's walk of their home bases, the 'site catchment areas' of Higgs and Vita-Finzi,[12] this was far from enough to satisfy their full requirements as human beings. In obtaining their most precious substances from remote environments the Danubian peasants were only following a pattern set long previously by Upper Palaeolithic man. The cave-dwellers of the Dordogne obtained shells from the Mediterranean and those of Mentone had apparently secured some of theirs from as far afield as the Indian Ocean.[13] It has been remarked that *Spondylus* shells resemble ivory and it is perhaps significant that ivory pegs with expanded terminals, possibly used for securing garments, were found together with shell ornaments in the Rhenish Flomborn cemetery.

In one important respect Neolithic farming communities broke new ground. They were the first to extend the realm of the precious to include substances of mineral origin. In the case of jade its aesthetic qualities only became apparent when the stone was polished. As a Chinese proverb has it:[14]

> If jade is not polished it cannot be made worth anything.
> If a man does not suffer trials he cannot be perfected.

The most likely context for discovering the qualities of jade would have been in the course of grinding and polishing axe or adze blades, one of the most widespread components of the Neolithic tool-kit from Europe to China. It is suggestive that axe or adze blades were the only artefacts common to ancient jade industries even including Mexico and New Zealand. The fact remains that the finest jades of the archaic period in China were symbolic in character and served no purpose in technology. This is not the occasion to discuss the process by which segmentally organized societies developed into vertically structured ones, though it is sure that significant clues will be found in the elaboration of the material symbols of emulation and status embodied in precious substances. For present purposes we may turn directly to societies in which the most significant relations were already vertically structured. Two points need to be emphasised. On the one hand the rise of polities headed by rulers and functioning by means of hierarchies of administrative, priestly and military officials, in itself called for graded insignia to legitimize the exercise of power. On the other the advances in technology and the emergence of increasingly specialized craftsmen, which characterized the economies of such societies, itself facilitated the creation and refinement of the necessary material symbols. This in turn made its impact on the choice of precious substances and not least on the skill and inventiveness with which these were mounted and displayed.

Although this applies to stratified societies as a class the relative values attached to particular substances varied in space and time. Whereas in Europe, the Nile Valley, south-west Asia and India the supreme measure of value attached to gold, in China and the Maya zone of Mesoamerica it went to jade. Gold was certainly used as a decorative inlay for bronzes as early as Shang times, but precious metals only began to be highly valued in China comparatively late and then as an outcome of influence from inner Asia and the west.[15] Indeed it is hardly too much to speak of jade and gold as embodying distinct standards of value. A classic confrontation between the two occurred when the Spaniard Diaz del Castillo was received by the Aztec leader Moctezuma in Mexico City. One may guess at the Spaniard's dismay when Moctezuma passed him some lumps of greenstone enjoining him to hand them to none other than his prince. The situation was only saved when the Aztec tactfully pointed out that each piece was worth two loads of gold.[16] Even today the Chinaman is as attached to jade as the Indian is to gold. Again, it is interesting to note that the Japanese, who never succumbed to jade or to any notable degree to gold, are now in their days of increasing wealth choosing platinum for their diamond rings.

Major changes occurred in the course of time in the availability both of different substances and of the techniques for shaping and displaying them. In each of the early civilizations the emphasis rested on opaque or translucent stones, in China on jade, but elsewhere on different forms of chalcedony, lapis lazuli and turquoise, the qualities of each of which could be displayed effectively by the simple process of polishing their surfaces smooth. When in the aftermath of Alexander the Great the Romans greatly increased the flow of trade between India and the Mediterranean they notably enlarged the range of precious substances. Those most coveted by the early civilizations of Egypt and Sumer, notably lapis lazuli, had been drawn from remote sources since Predynastic times. The improved trade routes of the Hellenistic period brought additional precious substances to the Mediterranean world, including emeralds, sapphires (and possibly rubies) and diamonds, as well as increased supplies of the Roman favourite, pearls. The coloured transparent stones could all be shown off by using the old technique of rubbing and polishing. Diamonds on the other hand were too hard, and during the Classical and Early Medieval periods could only be mounted as natural crystals, in which guise they served as symbols of hardness and strength, but not yet as signs of conspicuous wealth. The full potential of diamonds, as of their coloured counterparts, could be displayed only when Late Medieval lapidaries had learned the art of cutting them. It was only by directing light through facets at the right angles that craftsmen were able to release the fire and brilliance of diamonds and other transparent precious stones in a manner best fitted to symbolize and indeed embody the highest degree of excellence and majesty.

It remains to consider the third phase in the history of precious substances. What is the fate of such things in societies in which economic and political power has passed at least ostensibly from hierarchs to the general body of the population? How has jewellery fared in societies in which mass consumption prevails? And what of the notions of excellence of which it is a symbol? A leading British authority on jewellery ended her history[17] on an elegiac note:

It is not easy to see the future for the art of jewellery; it may even be considered that as an art it has not a future. For all the centuries of recorded time it has existed as an art in which style and fashion were set by the taste of an aristocracy; bourgeois jewellery, peasant jewellery in less precious materials such as we now call costume jewellery, all imitated court fashion . . . This source of inspiration has come to an end. Since 1918 comparatively few jewels have been designed for court wear in any country; since 1939 hardly any.

Since by definition jewellery reflects and in a sense embodies the social order, it should hardly be surprising, still less an occasion for despair, that social change as profound as that which marked the transition from an aristocratic to a democratic order should have brought about drastic changes in the uses to which precious substances were put. The emergence of more open societies in the western world had the effect of progressively enfranchising wider sections of the population, not merely politically and socially, but also as consumers. Whereas in hierarchically structured societies consumption of the most sought after goods was restricted to a relatively small class, in more democratic ones the mass of the population, no longer confined to the vicarious enjoyment of precious substances, was free to engage in their active consumption, unimpeded by sumptuary regulation, social disapproval or economic disability. There would only be reason for despair if more general affluence had been accompanied by a withdrawal from the regard for precious substances, since this would imply rejection of the notion of excellence, the basis not merely of cultural advance but of the very attainment of humanity.

This certainly did not happen. The problem has been rather how to satisfy an ever-increasing demand. As many progressive (but in reality reactionary) politicians have recently been finding to their cost, greater freedom of opportunity, so far from furthering equality, has only stimulated emulation and the appetite for symbols of success. The demand could be met only by increasing and widening supplies. The most direct way of doing so was by prospecting and directing capital to developing mines in new lands. Although gold was also sought for bullion and diamonds for industrial use, the expansion of the market for jewellery was one of the strongest impulses behind the booms in prospection and mining of the latter nineteenth and the twentieth centuries. Alternative ways of increasing supplies include the time-honoured one of debasement: one way of meeting the increased demand for gold jewellery was to increase the proportion of alloy by decreasing the carats of the metal used, to reduce the thickness of the metal or even to resort to gilding silver or base metal. In the case of precious stones the situation could be eased by using cheaper in place of more costly varieties – topaz for diamonds, tourmaline for emerald – and latterly by deploying the science and technology which brought the new social order into being to produce synthetic versions of the most precious varieties. Although the production of synthetic diamonds and rubies was only made possible by modern science, the idea of replacing natural by artificial substances in jewellery was one of remote antiquity. The ancient Egyptians had no compunction in substituting glass or faience for natural stones, the Byzantines explored the possibilities of enamel and the Chinese began to produce pearls by culture as early as the thirteenth century, a process which in the hands of the Japanese first brought fine pearls to the new mass market.

The fact that precious substances have been sucked with so much gusto into the stream of mass consumption admittedly has small bearing on jewellery as an art. Even at the upper end of the market the newly rich tended to choose the conventional settings for expensive stones. Yet Dr Evans might have taken some comfort from the wares offered to the more sophisticated sectors of the international market comprising elements from the former governing classes, leaders of financial and industrial corporations and not least the most successful popular entertainers. The most striking results have been achieved precisely by shedding conventional settings and mounting fine stones in such a way as to display their inherent qualities with the minimum of encroachment. The most skilled professional jewellers and their craftsmen have focussed attention on the qualities which first brought men to recognize particular substances as symbols of outstanding excellence.

2

◇◇

Organic materials

Most of the materials chiefly esteemed today for their symbolic value are of mineral origin. Yet in the small societies encountered during recent times by ethnologists in territories beyond those directly exploited by modern industrial economies they were predominantly organic and often highly perishable, including such things as furs and bird plumage. Archaeology suggests that this may also have been the case in prehistoric times. It is only the more durable, including some which still feature as objects of conspicuous consumption in the most sophisticated societies of today, that survive in the archaeological record. Notable among these are ivory, shells, rhinoceros horn, coral, amber and jet.

The organic substance most highly valued in recent times, pearls, will be treated more effectively alongside precious stones in Chapter 5. There is no evidence that pearls were sought or used as precious substances before the emergence of civilized states, and when pearls were adopted they were commonly used in jewellery together with precious stones.

Ivory

The term ivory[1] will be restricted, unless otherwise specified, to the material of elephant tusks. More frequently these derive from species still found in tropical and semi-tropical habitats, but fossil ivory from mammoths which flourished in the northern hemisphere during the Late Glacial period was also used. The term ivory has also been applied more loosely to a number of other substances. These include the curved canines and straight incisors of hippopotamus, walrus tusks, which project downwards from the upper jaw, the teeth of the sperm whale and the spirally twisted left tusk of the narwhal. Of these only hippopotamus teeth, which are denser and harder than ivory and of a gleaming whiteness, featured in any of the ancient civilizations. The animal survived throughout the dynastic period in Egypt. A delightful unguent jar from Mostagedda carved from ivory in the form of a hippopotamus argues that the animal attracted favourable attention. It also inspired deeper feelings. Amuletic wands were carved from its teeth with the object of deterring snakes and other noxious creatures.[2]

Walrus tusks were essentially substitutes for ivory in territories remote from supplies of elephant tusks. On the other hand in the far north they were evidently highly esteemed, to judge from the purposes to which they were put. Among the best known walrus ivories are the crozier from the Norse cathedral of St Nicholas at Gardar in south Greenland and the chessmen and board of Icelandic or Scottish workmanship from the Isle of Lewis, now in the British Museum.

The use of sperm whale teeth was apparently confined to Pacific islanders and Japanese mariners.

In commerce the designation ivory has sometimes been extended to nuts of the *Phytelephas* palm of South America, but this has only been used for making things like buttons, draughtsmen and netsuke.

True ivory is one of the few substances to have been highly esteemed for symbolic purposes wherever it could be obtained. One reason for its universal acceptance is that whereas noble metals and precious stones first rose to prominence in the context of civilized societies ivory had been recognized for much longer. A high regard for ivory has been part of the human heritage since the first appearance of modern man. The affluent clients of the smartest shops of Hong Kong, Paris, New York, London or Tokyo respond to the same qualities in ivory as those which attracted Palaeolithic mammoth hunters up to thirty thousand years ago and have continued to beguile all who have since had access to the material. Among its chief attributes are its pleasingly creamy colour, its smoothness and coolness to the touch, its fine grain and the comparative ease with which it can be wrought to a variety of precise shapes using such elementary techniques as sawing, rubbing, polishing and perforating. Ivory also lends itself readily to ornamentation. Designs can be executed by incision or drilling, relief carving can be done with quite simple tools and the comparative porosity of the material makes it relatively easy to achieve polychrome effects by staining. In civilized societies social considerations further intensified the value attached to ivory, as is the case of other materials. First its value was enhanced because, despite being an organic substance, ivory is remarkably durable. Second, in many communities where it was treasured at the most sophisticated level it was an exotic, even in some cases a mysterious substance, only to be secured from a distance. With this went, thirdly, that it was likely to be a royal monopoly available only by grace of the ruler who accumulated his stock of tusks through gifts, loot or tribute. It is a symptom of the high esteem in which ivory has consistently been held in civilized societies that it is often linked with gold and precious stones. Gold was used with ivory in the inlay of one of Tutankhamun's beds, and overlaid King Solomon's great ivory throne. Gold was also applied to ivory sculptures in Minoan Crete and in the chryselephantine work of Classical Greece.[3] A similarly close relation-

14

ship may be seen between ivory and some of the most important precious stones used in antiquity. For instance ivory was sometimes used in Sumer and the Aegean to make the seals more commonly made from different kinds of stone. Again, ivories themselves might sometimes be inset with precious stones. Some Assyrian ivories from Nimrud were studded with lapis lazuli, and the ivory wine cup buried with Fu Hao, consort of a Shang king, was inset with turquoise in early twelfth-century B.C. China. In sum the archaeological evidence confirms in rather a striking manner the impression conveyed by writers who held to the view that ivory connoted luxury and in itself constituted a treasure in much the same sense as gold, silver and precious stones.

One of the ways in which Upper Palaeolithic man differed from his predecessors lay in the use he made of the skeletal structures of his food animals.[4] The material recovered from his settlements displays a keen appreciation of the qualities of different parts of the skeleton and their potential uses. This makes it the more interesting that mammoth ivory, well adapted though it was by reason of its strength for tools and weapons, was reserved to a substantial extent for personal ornaments and sculptural representations. Why, when animal teeth were available and were in fact used for necklaces by the simple process of perforating their roots, did men go to the trouble to carve beads from solid ivory? Perhaps people who depended as much on mammoths as the hunting bands of south Russia or central Europe did during the late Ice Age sought by such means to identify with their gigantic and awesome prey. The source of ivory may well have been a factor in reserving it very largely to symbolic rather than practical use. As a Chinese proverb has it, 'ivory does not come from a rat's mouth'. The mammoth hunters of Mezine and Sunghir made use of sections of mammoth tusk to carve penannular bracelets perforated at either end to engage thongs which could be pulled tight when the wearer was actively engaged. The engraving on the outer face of the Mezine example, while decorative, is also likely to have had a symbolic meaning.

Another use of ivory initiated by Late Pleistocene hunting bands was for figurines. Most of these, including notable assemblages from the mammoth-hunter encampments of Gagarino and Kostienki I in the Ukraine and the head and torso from the French cave site of Brassempouy, depict unclothed women. (fig. 4). Men were rarely shown, but the figure from Brunn shows that the male form was sometimes featured. Among animals the graceful horse from Les Espélugues, Lourdes, and the mammoths from Predmost in Moravia are outstanding. If the purpose of these ivory carvings remains a matter for discussion, it is evident that Upper Palaeolithic man began the custom of using ivory as a medium for animal and human sculpture which has lasted down to modern times.

15

If northern peoples continued to draw occasionally on fossil ivory, by far the most significant sources open to ivory carvers since the emergence of civilized societies have been the tusks of living herds of elephants. Of the two species distinguished by zoologists, one is confined to Africa, the other to southern and south-eastern Asia, including India, Burma, Malaya, Indo-China, Sumatra and Ceylon. Archaeology suggests that their territory was once more extensive. Rock-engravings in the Sahara and across north Africa from Egypt to Morocco[5] argue that elephants extended, at least locally, as far north as the Mediterranean during the later Stone Age. There is even evidence to suggest that elephants existed as far as the Upper Euphrates basin as late as the first millennium B.C.[6] Syrian ivory contributed to the supplies available to New Kingdom Egypt and at the same time provided the material for the flourishing school of ivory workers

Figure 4 Ivory carving of a woman's head from Brassempouy, Landes, France. This head once formed part of a small figurine carved from mammoth ivory. It belongs to a group of 'Venus' figurines most of which date from an early phase of the Upper Palaeolithic found over a territory extending from the Pyrenees and North Italy to Central Europe and South Russia. Many specimens are made from mammoth ivory, but some were made from a variety of stones or even from fired clay. (Musée des Antiquités Nationales, St Germain-en-Laye)

16

based on Tyre, the products of which enriched the civilizations of the east Mediterranean and Assyria.

As with other precious substances, ivory was used far beyond the areas where it occurred in nature. Even the ancient Egyptians, as we have seen, welcomed gifts or exacted tribute from Syria to add to the resources of equatorial Africa. The Sumerians certainly maintained trade contacts with India, as we know from the testimony of seal-stones. That they may have obtained some of their supplies of ivory from that quarter is suggested by finds of this material at the intermediate entrepot of Bahrain. On the other hand the fine ivory work from Nimrud suggests that the Assyrians made use of the workshops of Tyre, which by this time may have been constrained to use ivory from Africa. To the peoples of the Aegean islands and mainland Greece ivory was all the more precious for being an exotic material.

The trade network which nourished Hellenistic and Roman civilization embraced both the main sources of ivory, and in the case of the Romans commerce was supplemented by loot. According to Livy, Scipio Africanus celebrated his triumph by parading over a thousand elephant tusks along with captured standards and a great quantity of gold and silver.[7] Ivory continued to serve many of the same purposes in Christendom as it did in Classical antiquity. The rulers of Byzantium drew at least some of their supplies from India. A graphic scene of an embassy from India arriving with elephant tusks and wild animals is depicted on the lower register of the Barberini Ivory (c. 500 A.D.) (fig. 5).[8] Ivory reached the west partly from Egypt but also, after the Arab conquerors of North Africa had opened up the trans-Saharan route from Tunisia by way of Chad to the Niger, directly from equatorial Africa. Ivory was certainly carried as far north as York during Anglo-Saxon times. Beyond that it was replaced by walrus tusk and even whale-bone.

The Chinese, who used ivory for elaborately carved handles and vessels as early as the Shang dynasty and in later times used it for a wide variety of personal items such as brush pots, wrist-rests, boxes, seals, snuff boxes and fans,[9] had increasingly to import the material as the elephant herds in the southern provinces diminished. It is possible that supplies may have reached them by the same inner Asian route by which they imported jade and turquoise and exported silk. It is significant that during the second and third centuries B.C. what was probably a customs post at Begram, Afghanistan,[10] on the route between China and the west was handling ivory as well as glass and bronzes from the west in addition to lacquer bowls from China. Yet unquestionably the Chinese drew their main supplies of ivory from the south. From Han times they were importing ivory along with rhinoceros horn, kingfisher feathers and pearls from North

Figure 5 The Barberini Ivory. The central register shows a Roman Emperor on horseback, perhaps Anastasius I (491–518), receiving (lower register) an embassy from India bearing elephant tusks. About A.D. 500. (Musée du Louvre)

Vietnam. By the Tang dynasty, when ivory was reaching the Chinese court in sufficient quantities to feature among the presents sent to the rulers of Japan, where they are still preserved in the Shoso-in treasure-house at Nara, African as well as Indian sources were being drawn upon. Arab traders exploited the winds of the Indian Ocean to knit together the Persian Gulf, the coast of east Africa and the Malabar coast of south India, where goods could be transferred from dhows to junks bound for Canton.[11] According to the Song geographer Chao-ju-kua it was this maritime network that brought to the Chinese several of the materials they held in highest regard next to jade.

The decorative potential of ivory and its pleasant feel were first explored by Upper Palaeolithic man, but more sophisticated uses of the material had to wait on the development of more complex societies marked by more or less pronounced hierarchies. A lively indication of the ways in which ivory could minister to the complex life-style of a young Egyptian ruler is provided by finds from the tomb of Tutankhamun who reigned during the mid-fourteenth century B.C.[12] Apart from furnishing armlets and bracelets ivory was used for the handles of the king's walking-stick, ostrich feather fan and whip. It also served for his ink and paint palettes, his gaming boards, and pieces, his caskets and his jewel-cases, one of the latter cut from solid ivory, fitted with knobs, hinges and covers for the feet of gold. The king's head-rest was also made of ivory and the same material served as an inlay to embellish his ebony bed, stools and chair. A similar degree of luxury and sophistication in the use of ivory was displayed by the upper levels of the partly contemporary societies of Syria, the east Mediterranean islands and mainland Greece.[13] Novel features included finely toothed combs, mirror handles, embellishments for chariots and horse harness and the use of ivory as an alternative to precious stones for cylinder and stamp seals. The Minoans were also adept at sculpting human figures from ivory, a practice they may have taken over from Egypt, where sensitive carvings of rulers had been a feature of Old Kingdom art.[14] The ivory figure of a Minoan deity grasping two snakes made of gold (fig. 6) is of interest as an essay in chryselephantine work combining the crafts of ivory carving and gold smithing.[15] The statues of Athena and Zeus respectively at Athens and Olympia by the sculptor Phidias carried forward the tradition, but the figures in which flesh was rendered in ivory and hair and garments in beaten gold[16] were in this case around forty feet high. Although during Roman times bronze and marble were the materials most commonly used in sculpture, the tradition of reproducing the human form in ivory was continued in the small scale reliefs carved on the backs of the writing tablets which served as official presents in pagan antiquity and as votive offerings in the early Christian church.[17] In Christian iconography ivory played a continuing role in sculpture as well as in low relief carvings. Some of the finest ivory carvings were later made to celebrate outstanding members of lay society. In this respect the Dieppe school was famous. Examples of the work of a leading exponent now in the British Museum include portrait busts of George I and Sir Isaac Newton by David Le Marchant (1674–1726).

Figure 6 Minoan snake goddess, Crete, c. 1600–1500 B.C. (Museum of Fine Arts, Boston).

Rhinoceros horn

Rhinoceros horns, in reality agglutinated masses of hair, are not attached directly to the skull but grow from the tough hide. Although the horn lent itself to delicate work and when finished had a smooth feel, its natural colour, yellow, mottled and streaked with grey, was so unattractive that the Chinese stained the objects they carved from it an artificial brown. Indeed rhinoceros horn is one of the few materials regarded as precious in the past not to appear inherently attractive to most modern observers. When King Gustav VI Adolph of Sweden[18] turned his attention to objects made of this material in building his collection of Chinese art there were those who found it difficult to reconcile such pieces with his choice porcelains, jades and lacquer-work. The king was determined and rightly so that his collection, now the property of his former subjects, should reflect Chinese rather than European taste. He was open-minded enough to appreciate the skill displayed in the carving and scholarly enough to value the insight into Chinese mentality offered by its imagery. The dragons, lilies, magnolias and lotus as well as the traditional activities and scenery depicted on the wine cups may be compared with those employed by jade-carvers and painters and deployed in poetry.

Figure 7 Rhinoceros horn cup, Ming dynasty, in the collection of H.M. King Gustav VI Adolf of Sweden (Östasiatiska Museet, Stockholm).

Although the Chinese transformed rhinoceros horn into forms of customary refinement, it seems unlikely that they went to the trouble of removing agglutinated masses of hair from rhinoceros snouts and lavishing such skill on them for purely aesthetic reasons. Indeed we know that they valued the material in part for its supposed aphrodisiac properties and in part because of its sensitivity to mercury, which helped to insure against drinking poisoned wine. It is an unhappy outcome of its aphrodisiac and prophylactic reputation that rhinoceros horn still commands a price high enough to threaten the very survival of the rhinoceros. Today the biggest threat stems from the prestige attaching to rhinoceros horn dagger handles among the men of North Yemen, who are apparently ready to pay several hundred pounds for an ornate dagger with a handle of this material.[19] Of the four genera recognized today, the one-horned *Rhinoceros* exists in India and Java and the three two-horned genera in Sumatra (*Dicerorhinus*) and Africa (the white *Cerafotherium* and the black *Diceros*). During the early dynasties a two-horned genus extended into the northern heartland of Chinese civilization. Lifelike representations of *Dicerorhinus* in the form of bronze wine containers (*zun*)[20] are known from the late Shang but also from as late as the Late Zou–Western Han period dating from the third century B.C. It was the transformation of the environment resulting from the spread of intensive agriculture that caused the rhinoceros, like the elephant, to retreat to the southernmost provinces of China and ultimately to abandon them completely. The animal survived locally in Szchewan, Kwangsi and Yunnan into the late Tang and early Song dynasties.[21] The earliest artefacts made from rhinoceros horn to survive are some plain wine cups in the Shoso-in collection at Nara dating from the Tang dynasty. It was precisely at this time that the Chinese were having to obtain supplies as tribute or by way of trade from the southernmost provinces of the empire or from Java and Sumatra. During the Tang period, when trade between the Persian Gulf, East Africa, India and the Chinese coast was conducted by Arab sea-farers, rhinoceros horn was among the chief imports to Canton.[22] Even larger supplies became available as the Chinese entered into direct trade with ports on the coast of east Africa. It is likely that many if not most of the elaborately carved wine cups of Ming and Qing date were in fact made of horn imported from Africa. As the material became more readily available it was used to meet the needs of the increasingly consumer-oriented cities of China as well as contributing to the export of finished objects. In addition to wine cups rhinoceros horn was used to make small sculptures of such traditional figures as Kuan Yin and Pu Tai, as well as a variety of desk furniture, toilet requisites and personal ornaments.

D Sumerian court jewellery

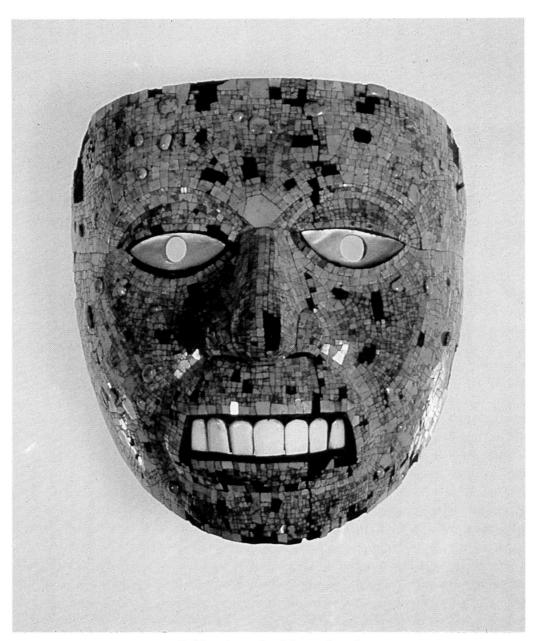

E Mosaic mask of Quetzalcoatl

Cowrie shells

Shells were among the first substances to be held in high regard despite being of very limited use for practical purposes. This is hardly surprising. Shells are pleasant to handle and regard, can easily be perforated to serve as ornaments and in addition are remarkably durable. The excavation of Upper Palaeolithic burials has shown that they were already being sewn onto clothing in positions which suggest that their role was symbolic as well as decorative. A crouched burial at Laugerie-Basse in the Dordogne had pairs of cowrie shells on each upper arm bone, two each on the forehead and either foot and four disposed around the knees and thighs. Perforated cowrie shells were also found on either thigh of an old man of the Upper Palaeolithic period buried in the Barma Grande, Mentone. A sign of the importance attached to shells at this time is that some were obtained from a distance. Those with the Laugerie-Basse burial were cowries (*Cyprea pyrum* Gmelin and *C. lurida* L.), both from the Mediterranean. More striking still, fragments of the shell of *Cassis rufa* from the Grotte des Enfants near Mentone came from as far away as the Indian Ocean.[23]

The attraction of shells has often been enhanced by attributing to them symbolic meanings suggested by more or less fanciful resemblances. This applies notably to cowrie shells which partly for this reason have maintained their status down to the ethnographic present in many parts of the world. The aperture of the cowrie shell into which the mollusc withdraws has suggested two quite different analogies, the vulva and the eye, the one promoting fertility, the other serving as a prophylactic against the evil eye and at the same time seeking to promote good sight. Because of their symbolic potency cowrie shells were often copied in precious metals. This was especially common during the Middle Kingdom in Egypt. From the twelfth dynasty one may cite a necklace or girdle made up of electrum cowries (Plate C) with beads of lapis lazuli, carnelian, amethyst and electrum, said to have come from Thebes,[24] or again one made up of gold cowries from Dashur, linked with the mother-goddess Hathor.[25] Ethnology shows that girdles of cowrie shells were worn by Tibetan women as charms against barrenness. The link between cowries and eyes is documented by archaeology as well as ethnology. The custom of inserting cowrie shells into eye sockets is at least ten thousand years old, as shown by the numerous skulls with plastered faces having cowrie shells inserted in the eye sockets from Natufian levels at Jericho.[26] To judge from the numerous clay copies of human skulls, plastered with clay and having cowrie shells in place of eyes, from the Sepik River, a similar practice obtained in New Guinea in recent times.[27] An alternative was to insert cowrie shells in the eye sockets of wooden images like those found as far apart as Togoland, the Philippines and New Zealand.[28] Ethnology documents other uses of cowries associating them with sight. For instance, the Pacific islanders use them to decorate the prows of their canoes and Malaysian fishermen attach cowries to their nets. The same basic idea explains why cowries were mounted on horse-harnesses in many lands, including India, Persia, Hungary, Japan, and China as early as the western Zhou dynasty.[29]

The fact that cowries were reproduced in gold and electrum has already been quoted as evidence of their high standing in ancient Egypt. Their occurrence with other prestigious grave goods in the burials of royal personages in ancient China points in the same direction. Nearly seven thousand were found with remains of the Lady Hao, consort of Wu Ding, fourth king of the Shang dynasty, together with sixteen human and six canine sacrifices, many bronzes including over two hundred ritual vessels, and large numbers of jades.[30] Another clue to the standing of cowries in ancient Chinese society is that the relevant hieroglyph features among symbols denoting 'precious', 'wealth', 'tribute' and 'currency'.[31] It is also significant that cowries featured, together with jades, spirits and weapons, among the gifts made on the occasion of official investitures.[32] Cowries of the *Monetaria moneta* species also served as currency in ancient China as well as in other parts of Asia and some areas of Africa. For this they were peculiarly well suited by reason of their durability, portability, uniformity and ease of recog-

Figure 8 Human skull from a Pre-pottery Neolithic level at Jericho, seventh millennium B.C. The face is plastered and cowrie shells are set in the eye sockets. (British School of Archaeology in Jerusalem)

nition. Cowries probably came into use as currency as early as the Shang dynasty. Although they began to be replaced in the more advanced parts of China by bronze coins by the middle of the first millennium B.C., they still circulated in some provinces at the time of Marco Polo's visit during the late thirteenth century. One of the chief limitations on cowries as currency is that their value was liable to be severely depressed by reason of the huge quantities in which they occur in some areas and the fact that they could so readily be transported by ship. Where they survived as currency down to modern times, as they did in west Africa and locally in south-east Asia, it was only as small change. At a time when copper coinage had already been in use for some centuries at the heart of the Han empire, wealthy leaders of the tien community in the southern province of Yunnan were storing hundreds of thousands of cowries in the great bronze drums which were a prominent feature of the Dong-Son culture focussed on northern Indo-China.[33] A possible source of these cowries would have been the

Figure 9 Clay copy of a human skull, plastered with clay and with inset cowries, from the Sepik River, New Guinea. (Museum für Volkerkunde, Berlin)

Pulo Condore islands off the Mekong delta of south Vietnam.[34] Beyond doubt the Maldive Islands in the Bay of Bengal were much the richest source in southern Asia at the time of the Arab and Portuguese domination of the Indian Ocean.[35] Arab traders carried the shells to India and westwards to the great entrepot of Zanzibar. There they were used to obtain ivory and rhinoceros horn for the Chinese market. A striking instance of the effect of trade is to be seen in West Africa, where cowries played a key role in cultural life, even though absent from local seas. Already in prehistoric times the Nok people of northern Nigeria were copying cowrie shells in their native tin-bearing cassiterite to make beads.[36] Supplies of cowrie reached West Africa in two main streams. The first were carried from the Maldive fisheries by Arab traders and reached West Africa by way of the Red Sea or Zanzibar, crossing the continent overland. The second came when the Portuguese, having rounded the Cape and carried western merchandise to India, south-east Asia and China, took on board quantities of cowries as ballast for the return voyage. These they unloaded at West African ports along with cloth, linen, glass beads and bronze manillas. In return they took on ivory and slaves to complete the round. For the West Africans this Portuguese traffic brought cowries in unexampled plenty. Although their sheer abundance depressed the value of cowries as units of currency, it provided a wealth of shells for decorative purposes as well as a medium for displaying the wealth and demonstrating the prestige of rulers. The floors and walls of the apartments reserved for the Oba of Benin were completely paved with imported shells.

Archaeology has even recovered cowrie money far away in central and northern Europe. Excavations at the Late Hallstatt cemetery of Donja Dolina on the Save revealed seven cowrie shell beads on a necklace made up of hundreds of amber beads and some made of coloured glass.[37] Further north cowries were found along with Byzantine and Kufic money at the Viking trading station of Birka on Björkö island in Lake Malar west of Stockholm, Sweden.[38]

Coral

The best known coral fisheries are those which have long been conducted in the warm waters of the west Mediterranean between North Africa and Italy, and around Malaysia and Japan. Once established in any region the fisheries tend to be long enduring since coral is a self-perpetuating resource. Its tree-like structures provide shelters for the polyps, the anemone-like organisms from whose secretions it is built. The industry which has flourished since Classical times at Torre del Greco south of Naples has developed such expertise that even corals of

Japanese origin are imported there for fabrication. Coral occurs in a wide range of colours. The kind most widely favoured for jewellery, at least in the west, was a delicate pale pink, but in and around Hawaii a black variety might be used for this purpose, and the Chinese were particularly keen on deep red coral for their carvings. In the Mediterranean coral was so abundant and so easily harvested that it never became valuable, at least in the home territory. It could be gathered by inshore divers down to ten metres and where necessary dredged from greater depths by nets weighed down by heavy timber frames. Its relative cheapness combined with its decorative qualities ensured widespread local use for popular jewellery. Where it was introduced as an exotic substance it was more highly valued. Among the Celtic peoples of temperate Europe it was applied to La Tène bronze safety-pins and torcs[39] as well as to prestige objects such as bronze wine flagons. Its status was even higher when it was carried to West Africa by Portuguese traders in 1522 and subsequently by Dutch and French ships. The merchants realised how well pink coral would show up against dark pigmentation. In presenting it direct to the kings of Benin they showed good understanding.[40] The rulers promptly monopolized it for their own regalia and as a medium for bestowing honour and obligations on their retainers. When the French Captain Landolphe delivered a consignment of coral necklaces and rosaries in 1778, he and his companions were invited to inspect the royal regalia carried in procession by the Oba and his chiefs to a shrine in one of the palace courtyards. There they witnessed the blood of sacrifices, human as well as animal, being scattered over ornaments including a ceremonial collar of fifteen to twenty coral necklaces. Not much more than a century later the king's successor made his submission to the British after the punitive raid on Benin City. He did so wearing his coral regalia. He was

simply covered with masses of strings of coral, interspersed with large pieces . . . His head dress, which was in the shape of a Leghorn straw hat, was composed wholly of coral of excellent quality, meshed closely together, and must have weighed very heavily on his head, for it was constantly being temporarily removed by an attendant. His wrists up to his elbows were closely covered with coral bangles, so were his ankles . . . his breast was completely hidden from view by the coral beads encircling his neck.[41]

Wearing coral was a royal privilege which the king was able to delegate to his retainers. At his ceremonial submission to the British his wives wore coral necklaces and other ornaments. The French Captain Landolphe had previously noted that leading members of the Oba's councils wore two necklaces as well as anklets and bracelets of coral, whereas lesser untitled officials were allowed only one necklace each. Apart from their role in maintaining good relations between a ruler and his chief associates, coral beads, like cowrie shells, served as media of exchange. An ounce of coral beads was reckoned to be worth ten large jars of oil.[42]

Amber

Although amber can no longer be regarded as among the more precious substances, it was certainly rated as precious in former times.[43] During the Bronze Age it ranked with gold in Middle Europe and Mycenaean Greece. Again, in Ming China its standing was made evident by its use for insetting silver hair-pins and filigree head-dresses.[44] Amber has always been used predominantly for jewellery, but in later times it also served to meet a variety of personal needs such as smoking gear, rosaries and worry beads. Its high status is reflected in the fact that though useless for industry it was nevertheless extensively traded both in Europe and the Far East. During the historical period East Prussian amber was considered sufficiently valuable to be declared Crown property, its exploitation subject to licence. It was also considered worthy of inclusion among the gifts presented to Tsars by envoys from the Electors of Brandenburg. The Tsars for their part had an amber room built at their summer residence of Tsarskoe Selo to house the collection in the same way as rulers of the period had their porcelain rooms. More surprisingly, the Pitti Palace at Florence housed a collection of amber vessels, cabinets, figures, caskets and crucifixes.

The prime attraction of amber rests on its smoothness, its ease of handling and its colour, ranging from a transparent varnish-like effect to a deep brown or cloudy yellow colour. In recent times the market reflected strong regional preferences. Whereas the Chinese would take only clear amber and the Dutch had a preference for it, the Russians went for cloudy amber. The fact that many people have attributed therapeutic properties to amber probably stems from its electrical qualities – for instance, when rubbed it will attract straw or paper. Again it has enjoyed a certain popularity as an incense. It burns slowly when ignited and gives off a pleasing smell.

Many fanciful explanations have been advanced to account for amber, but the Elder Pliny already diagnosed it as a fossil resin of pine and the same opinion was also entertained in China before the Tang period. Two main kinds of European amber have been distinguished on chemical grounds. One, marked by relatively high values for succinic acid, is found predominantly in northern Europe, though it also occurs with lower succinic acid contents in Miocene formations exposed on the banks of the river Bazeu in Rumania. The other kind, lacking succinic acid but marked by the presence of organic sulphur, is at home in the Mediterranean.

Known after the Sicilian river Simeto as Simetite, this was dark red and translucent, sometimes with a blue or green fluorescence.

In modern times most of the succinic amber from the Baltic is won by mining a 15–18 m thick layer of greensand under a bed of lignite overlaid by superficial layers of marl and sand. The basal 1.5 metres of the greensand layer can be immensely prolific. It is from this stratum, extending under the sea, that the amber found on the shore gets eroded. Amber can be collected on the south shores of the Baltic from Lithuania to Denmark, as well as of the North Sea from west Jutland, north-west Germany and south-east England.

Archaeology shows that the material was appreciated by the local population of this region long before they had begun to practise farming. The recovery of a piece engraved with a horse's head from the reindeer-hunters' station of Meiendorf, north-east of Hamburg, shows indeed that this began during the final stage of the Ice Age.[45] Finds are commoner from Mesolithic sites on either side of the North Sea, including Star Carr,[46] Yorkshire, and the Danish sites of Svaerdborg, Lundby and Verup, not to mention stray finds including figures of a bear, an elk head and neck and numerous beads and amulets (Plate A).[47] It is also worthy of note that East Prussian amber was being traded among hunter-fisher communities over extensive tracts of Finland, the East Baltic countries and north-east Russia as far as Lake Onega and the Upper Volga,[48] as well as up and down the coast of western Norway,[49] at a time when Neolithic farming communities were being established in south Scandinavia. The Eskimos of Alaska were collecting amber and trading it at the late summer fair held at Kotzebue on the north-west coast down to recent times.[50]

The introduction of farming to the south Baltic region appears to have stimulated the use of amber ornaments.[51] Early neolithic farmers favoured multistrand crescentic necklaces made up of tubular and disc-shaped beads kept apart by laterally perforated oblong spacer beads and pulled together by triangular toggles. Their recovery from graves shows that they were used as personal ornaments, but the discovery of large numbers in the peat bogs, sometimes enclosed in eared flasks, suggests that they were also used as votive offerings, a sign which suggests in itself that amber was regarded as precious enough to serve as conspicuous waste. Its use for ornaments meant that it was subject to fairly rapid changes of fashion. The Middle Neolithic saw the introduction of axe and double-axe forms, but the battle-axe people themselves preferred perforated disc forms.

A new phase began when Baltic amber came to be used primarily for export in return for copper and tin. This most probably stemmed from the fact that the bronze-using Aunjetitz communities of Bohemia acquired a taste for amber from their Neolithic neighbours in Poland, defined in the archaeological record by their globular amphorae.[52] A taste for wearing amber was widely taken up among allied groups in south Germany and north Italy, at the head of the Adriatic.[53] The fact that it spread as far as the Mycenaean chiefdoms of Greece has sometimes been regarded as surprising. Yet it accords with the general rule that the more stratified societies become the greater their appetite for precious substances and the wider the range over which they were prepared to obtain supplies. The lords of Kakovatos and Mycenae[54] attracted gold, lapis lazuli and ivory, as well as amber, among the substances they used to advertise their status. By the time amber had reached the head of the Adriatic it was readily accessible by sea and it is hardly surprising that Kakovatos on the west coast of the Peloponnese should have yielded the richest find of amber from this period in Greece. A notable feature of the ambers from Kakovatos is that they include spacer beads with convergent perforations like those on similar pieces from tumulus burials in south Germany,[55] on the route from the amber coast of west Jutland to the Aegean world.

The power of stratified societies to attract rare and precious substances from afar is well exemplified by China. By comparison with jade, ivory or even rhinoceros horn, amber never featured prominently in China. It is nevertheless of interest that supplies were being drawn into the Chinese market as early as the Han empire, that it featured among the tribute rendered by the Turkish tribes of central Asia during the tenth century A.D., that the Portuguese introduced European amber through Macao and that during the Qing dynasty supplies were assured from the mines of north Burma, situated in the same region as the sources of jadeite. Needless to add, the Chinese were adept at imitating amber using such materials as copal, shellac and colophony.[56]

Jet

Jet, a mineralized vegetable product allied to lignite and cannel coal, exists in a variety of grades.[57] The most highly esteemed in Britain occurs in lumpy masses in the shale beds of the Upper Lias in north-east Yorkshire, but lower grades which were also used in prehistoric times occur in the Lower Lias of Dorset, Somerset, Gloucestershire and Shropshire. Analogous materials occur in other parts of the world. The very word from which jet has been derived, the Latin *gargetas*, is itself named after the river and town of Gagas or Gages in Lycia, Asia Minor. Again, one of the main sources of supply today and one that helped to put the Whitby industry out of business during the late nineteenth century, is centred on Oviedo in Spain, which draws its supplies from Lower Cretaceous marls in Asturias. The highest grade of jet typified by that from Whitby attracts by its dense, velvety black colour and the resinous lustre which can readily be enhanced by polishing. Its appeal is furthered by the fact that it can be electrified by rubbing, a property which has sometimes led to its being referred to as 'black amber'.

The jet industry based on north-east Yorkshire and in historical times focussed on Whitby was active from late Neolithic to Victorian times, a span of over four thousand years.[58] Throughout almost the whole of this period the industry depended on supplies eroded by the sea and cast up on shore. It was only as shore jet ceased to meet increasing demands that recourse was had to outcrops exposed on hillsides in the interior. The unpredictable occurrence of jet in the shale beds and the cost of constructing adits against a doubtful return meant that Whitby was overtaken during the final quarter of the nineteenth century by imports from Oviedo. The decline of the natural jet industry was hastened by the invention of an artificial substitute, ebonite or vulcanite, made by exposing india-rubber and sulphur to intense heat. Its final decline was brought about by a change in the public attitude to death. The custom of wearing jet jewellery as a symbol of mourning was quietly but effectively dropped.

No precise information about the extent of trade in jet can be expected until different varieties of jet have been more precisely defined. The probability is that beads from English Neolithic long barrows and enclosures and described as made of shale or inferior jet came from the Kimmeridge locality in Dorset where bracelets are known to have been made during the Early Iron Age.[59] Oblong sliders or belt fasteners from Late Neolithic contexts on the other hand resemble Yorkshire jet,[60] as do the conical buttons with V-perforations at the base associated with Beaker pottery and the crescentic multistrand necklaces with toggles and spacer beads dating from the Early Bronze Age (fig. 10).[61] The high esteem in which jet was held at this time is shown by its relation to amber and gold. Buttons and multistrand crescentic necklaces were made of amber as well as jet. As to gold, sheets of the metal were applied to the large shale buttons found with prestigious burials under a group of round barrows in Wessex.[62] Again, some of the decorative motifs applied to contemporary moon-shaped lunulae made from thinly hammered gold strongly suggest, notwithstanding a recent suggestion to the contrary, the arrangement of crescentic necklaces of jet or amber.[63] Jet ornaments continued to be buried with the dead during the Late Bronze Age and Early Iron Age in Yorkshire. Whitby jet was widely used in Roman Britain, and jet beads occur commonly in Anglian cemeteries of the Anglo-Saxon period. From

Viking times the strong settlement of the Whitby region finds a counterpart in exports of jet ornaments to west Norway. In the same context it is not surprising that jet should have been found together with other exotic imports including cowrie shells at the trading township of Birka on the island of Björkö, middle Sweden.[64]

Figure 10 Jet necklace of the early Bronze Age from Poltalloch, Argyllshire (National Museum of Antiquities of Scotland, Edinburgh)

3

<hr />

Jade

Archaeology and history combine to show that jade was rated more highly than gold among some of the most sophisticated and highly civilized peoples of antiquity both in the Old World and the New.[1] To understand the appeal of jade it is essential to handle the material itself. Only so can one appreciate the tactile as well as the visual qualities that render it attractive. Among these are its smoothness and coolness to the touch, a degree of translucency, and colours ranging from the white most highly esteemed by the Chinese to the various shades of green, yellow, orange, red, blue, mauve and black imparted by mineral components. It was such qualities, combined with its outstanding durability and scarcity, that led the Chinese to invest it with the symbolic qualities and applications that served to mark it out all the more emphatically as for them precious beyond all other substances.

As is well known, the word jade was an outcome of the medical qualities attributed to it by the Aztecs.[2] The Spaniards when they conquered Mexico were impressed by the effectiveness attributed to it by the indigenous population for treating pains in the sides and kidneys. This caused their own physicians to refer to the Aztec *chalchihuitl* as *piedra de yjada* or 'stone of the loin'. From this it was only a step to the French and English abbreviation 'jade'. The small quantities of Mexican jade imported into Spain were followed, in the wake of imports of porcelain, by accessions of jade pieces from China. It was the contrast between Mexican and Chinese jades that stimulated European scientists to investigate their respective compositions. The French mineralogist Alexis Damour distinguished two main groups, the amphibole rocks represented by nephrite and the pyroxenes which included jadeite and the little used chloromelanite. The two main groups differ in colour, and in the hand nephrite has an oilier feel than the glassier jadeite. All three are mainly composed of silica, but analysis reveals significant differences in other components.[3] Whereas nephrite contains a high proportion of magnesia and a considerable one of lime, neither of these is present except as traces in jadeite. By contrast jadeite contains plenty of alumina and soda, only weakly present in nephrite. Iron is present in each as traces, but even so provides useful markers. In nephrite it occurs exclusively as ferrous and in

33

jadeite as ferric oxide. In chloromelanite, whose dark colour goes with a strong representation of iron, this metal occurs in both forms. If today we follow scientific analysis, the peoples of the non-industrial world discriminated on the basis of sensory perception. Ethnologists have recorded that the New Zealand Maori recognized nearly a score of different kinds of jade, based on their perception of differing variations and combinations of colour and texture.[4] Conversely, where, as in the case of China, sources were remote, the type of jade most commonly used at any particular juncture depended on what happened to be most readily accessible at the time.

In the course of history jade has been held in highest esteem among four quite distinct groups of people, the Chinese from the later stone age to modern times, the neolithic inhabitants of substantial parts of temperate Europe, the Maori of New Zealand and in the New World the inhabitants of early Mexico from as far back as Maya times. In the first instance the archaeological evidence is richly reinforced by written sources and modern usage and in the last two by the testimony of European explorers and colonists.

A point to emphasize is that in each case the material was obtained from natural sources distant and sometimes remote from the places where it was put to social use. Indeed both in Europe and in Mexico archaeologists encountered jade artefacts before knowing where the material came from. One of the classic confrontations of nineteenth-century ethnology stemmed from this very circumstance. Heinrich Fischer of Freiburg sought to explain the occurrence of jade objects in each of these areas as the outcome of trade or even of migration from the Far East.[5] A. B. Meyer of Leipzig preferred to think that if sufficiently diligent search was made natural sources of jade would be found in Europe itself.[6] As things turned out Meyer had the satisfaction of being able to point to alluvial sources of nephrite in eastern Switzerland and central Germany. The main drawback was that apart from in Switzerland itself nephrite was seldom used in prehistoric Europe. For instance, thirteen out of every fourteen jade axes from the British Isles were made of jadeite, and this was the predominant material used for the same purpose in France and Germany. Sources of jadeite have been particularly keenly sought in the British Isles and France,[7] but without result. The only sources so far identified in Europe comprise a few boulders from Piedmont and Switzerland. The rarity of finds of natural occurrences suggests how intensively the prehistoric Europeans must have sought rare substances and how much trouble they were prepared to go to in disseminating them.

Conclusive evidence that the sources of jade exploited in early Mexico were indigenous, like the economy and culture of the region itself, has been demonstrated only in comparatively recent times. Mineralogical studies have revealed rich sources of jadeite in the valley of the Montagua river which flows through southern Guatemala to the Gulf of Honduras. An additional source has now been traced to the Balsas river which flows south from the highlands of central Mexico to the Pacific.[8]

A notable feature of the greenstones which have played such a crucial role in the social life of the New Zealand Maori is that their sources were confined to the South Island, whereas due to the warmer climate and the requirements of important food plants, population was mainly concentrated on the North Island. The main sources of the keenly sought after nephrite are the gravels of the Arahura and Taramakau rivers in the northern half of the province of Westland. Other prized greenstones include an opaque nephrite from the river Dart and a translucent serpentine from Anita Bay, Milford Sound, both still further south.[9]

China

Although the Chinese have held jade in the highest regard for a matter of five thousand years and carried its symbolic use to levels of sophistication far beyond that of other peoples, no natural deposits are yet known from within the ancient limits of their country.[10] Until they supplemented it by jadeite from Burma during the eighteenth century, the Chinese relied on nephrite. Of this they had two main sources. By far the more important comprised rivers flowing down from the Kunlun Mountains in the regions of Khotan and Yarkand on the southeastern margin of the great basin containing the Taklamakan Desert. The distances over which jade was transported were enormous (fig. 11). It would take a radius of c. 3,600 km from this source to encompass the zones of China in which jade was being worked as far back as neolithic times. Yet it is worth remembering that the line of movement skirting the margin of the Taklamakan Desert is amply documented in Chinese history. Traffic was passing that way as early as the Shang dynasty and under the Han the silk route was crucial enough to be incorporated in the Empire. An independent hint that it was being used even earlier than the Shang dynasty is the appearance of turquoise, the nearest sources of which were situated on the Iranian plateau,[11] during the initial Erh-li-t'ou stage of the Bronze Age.[12] Turquoise was also present among the grave goods of Fu Hao, consort of a Shang king, in a tomb dating from the first half of the twelfth century B.C.[13] One of several nephrite halberd blades was hafted in a bronze mount studded with turquoise inlay. Another clue lies in the appearance among the jade figurines of a representation of an elephant, and the inclusion among the grave goods of a wine cup carved from an elephant tusk. It is true that elephants existed at that time in the extreme south of China, but the fact that Fu Hao's cup was studded with turquoise points to India as a possible source. Recent research on the Lungshan culture of eastern China[14] suggests that Shang jade technology was rooted in Neolithic antecedents. Without discounting the possibility that sources may yet be found in the mountains of China, present indications are that jade was already reaching China in Neolithic times by the route followed during the Bronze Age and down to the time of Sir Aurel Stein's travels in inner Asia at the turn of the nineteenth and twentieth centuries.[15]

The only other significant source of nephrite within reach was situated in and around the Vostochny Sayan range of inner Siberia, west of the southern end of Lake Baikal.[16] White nephrite rings were already being distributed, presumably from this source, as far west as Seima near Gorki at the confluence of the Oka and the Volga at a period broadly contemporary with the Shang dynasty. Yet there is no evidence that jade from this source was reaching China before the eighteenth century. When it did so, it was commonly in the guise of Siberian

Figure 11 The main sources of jade used in China.

spinach jade, a variety deep green in colour, flecked with black. Although this made less appeal to Chinese taste than paler varieties, it was found useful for feeding a growing export market to the west.

Jadeite, highly prized in modern China, did not appear there until well on in the eighteenth century. To begin with it came exclusively in the form of pebbles and boulders from the Kachin hills of north Burma. Substantial supplies had to wait on the mining of reefs first found as late as 1880 outcropping on the Tawmaw plateau.[17] The only other source of jadeite known from Asia was a relatively insignificant one in Japan. This was used to make comma shaped charms (*magatama*) of the kind placed with burials of the Tomb Period during the first half of the first millennium A.D. both in Japan itself and in Korea.[18] The most striking fact about the Japanese is that apart from this their attitude to jade was negative. Despite many cultural imports from China during their subsequent history, the islanders conspicuously failed to acquire a reverence for jade.

Wherever jade was used in early times it was derived from alluvial sources. Only later was the material won by mining. By concentrating on pebbles carried down from alpine sources in the beds of rivers and streams, full advantage could be taken of the erosive power of water in wearing away adhering rock and concentrating the sought after greenstone. Stream beds offered a further bonus. When Captain Cook visited the nephrite zone of the South Island of New Zealand, the Maori described jade as a fish that became stone only when brought to land.[19] Anyone who has paddled in the shallow waters of the Arahura river will understand how such a concept arose. Nephrite pebbles show up by their colour and translucency much more clearly when viewed through flowing water than they do when dry. When the Jesuit Benedict Goës passed by Khotan in 1603 he reported that the superior kind of jade was taken from rivers 'almost in the same way in which divers fish for gems'.[20] According to a Chinese source originally published in 1637, women were made to strip and wade in streams to search for jade pebbles.[21] Around 1900 Sir Aurel Stein observed that 'the ancient industry of "fishing" for jade in the river bed after the summer floods still continues all along the valley'.[22] One of the drawbacks of alluvial sources is that under the pressure of a growing demand they tend to be depleted to a point at which more intensive methods are called for. In Stein's day the jade searchers of Chinese Turkestan had already taken to a kind of superficial mining, even if this amounted to no more than burrowing through overlying deposits to search the cobble bed down to depths of up to some twenty feet. A final stage involved tackling the native rock. One way of doing so, reported by Benedict Goës during the seventeenth century in Turkestan, was to apply heat and then split the rock by douching with cold water. A drawback of this fire-setting technique is that it was liable to impair the value of the product by cracking it. It is certain from the size of some Chinese jade artefacts that larger blocks were being quarried, probably in the region of Yarkand, from the sixteenth century.[23]

If the rarity of jade and the distances from which men were prepared to obtain supplies were one indication of the high regard in which the material was held by certain peoples, the investment in time and labour needed to shape this hard material to social needs was even more eloquent of its honorific status. On the Mohs scale for measuring the relative susceptibility to scratching of different minerals nephrite is rated at c. 6.5 by comparison with jadeite at 6.75, steel at 6 and diamond at 10. Although nephrite is less hard than jadeite on this scale, the tight felting of its fibrous crystals makes it exceptionally tough. As a result no kind of jade can be worked by cutting, flaking or pecking. The only way to shape it is to abrade it with even harder substances. A simple way of doing this was to use sandstone saws like those recently employed by the Alaskan Eskimos. An alternative was to apply the abrasive in powdered form to saws of materials like bone, wood, bamboo or iron. The initial task of obtaining blanks was performed by cutting parallel grooves sawn into the parent block. The main cost was time. It took a Maori around four weeks to extract a blank for an adze-blade from a pebble of nephrite and another six to shape it (fig. 12). Intermittent

Figure 12 Adze blade carved from a nephrite boulder. The Maori obtained their nephrite adze blades by cutting grooves from either face of a boulder, using abrasives and water. This was a laborious process and probably the equivalent of a week's hard work had been expended on this stone before it was abandoned. (Canterbury Museum, New Zealand)

F Maori *tiki* of greenstone

G Romanesque crozier of walrus ivory

polishing might continue almost indefinitely. Even in China where the working of jade was carried to the highest pitch of sophistication the lapidaries continued to rely on abrasion for working their material.[24] The only scope for progress was to find harder abrasives, devise more effective tools for applying them and not least to be prepared to invest more time, something which depended on more powerful patronage.

At the technical level the Chinese continued to rely on quartz sand, the abrasive used in each of the other centres of jade-working, up to the Tang dynasty, when it was reinforced by crushed garnet (Mohs 7.5). Corundum (Mohs 9), a crystallized mineral of the ruby group, was taken into use during the Song dynasty. The artificial abrasive carborundum, made in Sweden, the U.S.A. and Japan, was probably not available to the Chinese until after the war of 1914–18, since when its main use has been to impart a high surface gloss to finished articles. The hardest of all, diamonds (Mohs 10), probably reached China from India by the third century A.D., but their use was mainly confined to drilling fine perforations.

As to apparatus, saws were evidently effective enough to cut nephrite into the relatively thin sheets from which many archaic jades were made,[25] and bow-drills must have been available from the beginning. The introduction of iron and steel for implements, including saws, cutting discs, grinders, gouges and drills, undoubtedly enhanced the jade-worker's control over his material. In particular it eased the task of hollowing vessels, executing decorative designs in high relief and undertaking such *tours de force* as making vessels with pendant rings, sometimes with lids attached by chains, all carved from the same block of nephrite.[26] A key device was the rotary lathe operated by a foot treadle, which left the hands free. Lathes were introduced to China perhaps as early as the Ming dynasty and were still being used in the jade workshops of Peking when Howard Hansford came to study them in 1938–9.

The scarcity of jade, the distances from which supplies were drawn and the high input of labour needed to work it made it too expensive for ordinary working purposes. This was already the case during the archaic phase of Chinese jade-working.[27] No practical purpose has ever been attributed to the four-sided blocks with tubular perforations and grooved corners (*zong*) (fig. 13) that appeared already in the Neolithic culture of east China,[28] and the same applies to the animal amulets of Shang times. Even when the forms of such things as halberds or hoe-blades were rendered in jade they were too delicate for functional use. In China jade was from the first shaped for symbolic rather than industrial ends. Precisely what particular forms symbolized is often in doubt. Perforated discs (*bi*) (fig. 14) and the afore-mentioned *zong* came to designate

Figure 13 Ceremonial jade *zong* from south-east China; Neolithic c. 2500–2000 B.C. Height 203 mm. (British Museum)

heaven and earth, but it is not known whether this was their original meaning.[29] Similarly, the interpretation of barbed and serrated discs (*hsuan chi*) as astronomical devices intended to give guidance in scheduling affairs[30] is only the most plausible of those already advanced. On the other hand there is conclusive evidence that by the Shang dynasty jade was being used to denote high rank. This is established by excavated burials. The undisturbed burial of Fu Hao at An-yang featured the sacrifice of sixteen people, six dogs and fifteen hundred objects,

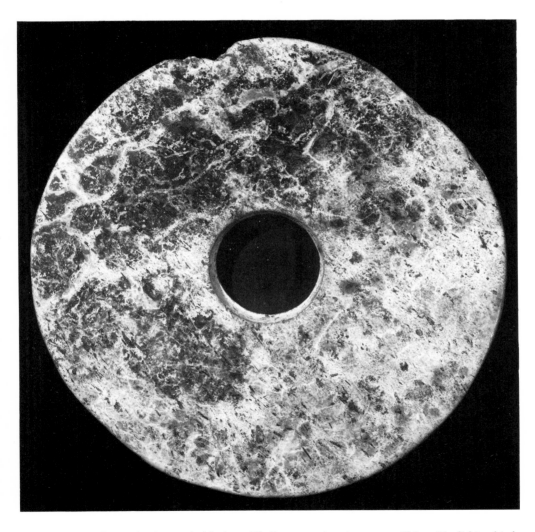

Figure 14 Hardstone *bi*-ring probably from Zhejiang province in eastern China. Neolithic, third millennium B.C. Diameter 190 mm. (British Museum)

including no less than five hundred jades.[31] An interesting feature of the jades is that they included numerous pieces in the form of halberd blades. Their delicate character and the fact that they accompanied a woman means that they can hardly be interpreted literally as weapons. They can only have been insignia marking the lady's status as a royal consort.

The association of jade with rulers in China is documented in many ways. Jade accessories featured in rituals designed to ensure good relations between emperors and the celestial powers on whom they depended to ensure the continued mandate of heaven. As we have seen in respect of Fu Hao, a ruler's association with jade extended to his immediate family. An emperor could also delegate authority to his officials through the medium of jade symbols. The passing of a jade emblem served not merely to denote but to validate authority. In like wise the exchange of jades on such auspicious occasions as the emperor's birthday served to consolidate mutual relations between ruler and officials. Jades also made acceptable sacrifices to mark the sealing of treaties or the completion of other transactions of moment. The pre-eminent virtue attached to jade contributed in a variety of ways to the smooth working of the kind of hierarchical society reflected in the works of Confucius (551–479 B.C.).

One of the qualities of jade that contributed most to social harmony was the musical nature of the notes it emitted when struck. Tomb finds suggest that this quality was appreciated as early as the Shang dynasty. Musical jades were angular in form, resembling carpenters' squares. They were mounted through perforations to hang from wooden frames, singly as gongs and in sets of sixteen of graduated thickness for chimes.[32] Struck at such moments as the end of a verse, these enhanced the impact of rituals performed to promote harmony and enhance the dignity of the state over a period of at least three millennia from the Shang to the Qing dynasties.

Emphasis has rested so far on the public role of jade in furthering the harmonious working of Chinese society by cementing relations between the ruler, heaven and the several levels of the bureaucratic hierarchy by which the empire was administered. Jade also played a more private role in meeting the anxieties and promoting the pleasure of individuals. The deposit of a variety of jade amulets for ensuring good fortune and warding off ill-luck among Fu Hao's grave goods shows that jade had already begun to be directed to individual needs as early as the Shang dynasty, at least for one at the peak of the social system. The amulets, including flat plates designed for attachment to a background material as well as three-dimensional figurines, featured a wide range of animals identified in mythology with attributes favourable to human well-being. Fu Hao's charms included birds, fish and mammals. Birds were held in high regard because of their ability to soar between heaven and earth, among them birds of prey valued as protective spirits because of their propensity to kill the small animals that damage crops, as well as phoenixes considered to be especially effective against the scourges of drought and fire. Fish were included above all as symbols of fertility and plenty. Among the numerous animals represented were bears, tigers, elephants, tortoises and not least dragons, inhabitants of clouds and bringers of rain.

Although the personal concerns of rulers and their consorts began to find reflection in the jade record as early as the Shang dynasty, in general state affairs were dominant at this time. The shift from almost total preoccupation with ritual, cosmological and hierarchical affairs came in the context of political turmoil and spiritual conflict during the Late (Eastern) Zhou period.[33] This was duly reflected in the jade inventory. Among innovations to appear at this juncture were such personal items as cups and tumblers, belt or girdle hooks, sword pommels and scabbard fittings like buckles and chapes, as well as bowmen's rings.

During the Han period expression was given to the most vital of human concerns. The durability of jade alone encouraged the idea that it might prolong life or at least, by protecting the body from decay after death, ensure the possibility of resurrection. Chinese alchemists also believed that longevity could be encouraged by eating jade. This could be done by mixing powdered jade with more palatable foods. A simpler and more pleasant way of achieving the same end was to use jade vessels for food and drink. According to an early eighteenth-century edition of an early Han source,[34] it was reported of the emperor Wen that he

Figure 15 Hawk pendant from tomb 5 at An-yang (c. 1300–1030 B.C.). Rubbing to show decoration; height 62 mm.

'acquired a drinking cup of jade, on which was carved this inscription: "Master of Mankind, may Thy Life be prolonged to the great delight of this world!" '. In respect of the dead the idea was to prevent decay of the body. One way of attempting this was to stuff the mouth with rice along with precious substances consonant with the rank of the deceased. Jade was apparently right for an emperor, jade and pearls for an official of grades 1–3, jade and gold for those of grades 4–7, gold or silver for gentry and silver for ordinary folk.[35] The practice of placing a jade representation of a cicada on the tongue of the deceased probably stemmed from a knowledge of the life history of this insect, which re-emerges after spending a period underground.[36] Ways of arresting decay also included covering eyes and mouth with carefully shaped pieces of jade and inserting jade plugs into other orifices of the body.[37] The climax was reached in royal burials of the Han dynasty like those of Liu Sheng, Prince Qing of Qungshan, and his wife Dou Wan, which featured prominently at the Burlington House, London, exhibition of 1975. The bodies of these two were encased in garments each made up of over two thousand oblong plates of jade knit together by metal threads (fig. 16). The plates themselves were of standard design, but a nice sense of hierarchy was observed in the kind of metal used to tie them together – gold for the prince, silver for his wife and bronze for concubines.[38]

The increasing growth of bureaucracy in Chinese society under later dynasties was reflected in the multiplication of jade artefacts to meet the particular needs of scholar administrators.[39] Square seals, often surmounted by dragons, and

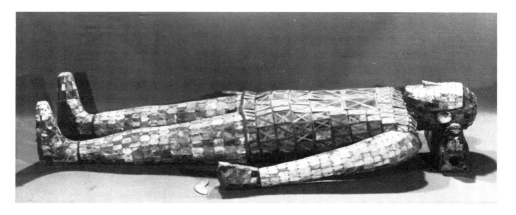

Figure 16 Jade burial garment of the Han pricess Dou Wan from Mancheng in Hebei province, late second century B.C. The garment is made up of 2,156 saw-cut jade plates joined together by metal threads knotted through the corners. It would probably have taken an expert jadesmith more than ten years to complete such a suit.

boxes to hold the necessary wax satisfied a basic requirement of a high official at his desk. Other accessories included those associated with calligraphy and painting, notably jade brush handles, arm rests to steady the hand, ink-stones, brush and water pots, paper-weights and, not least, elaborately carved screens pleasing to contemplate and designed to prevent ink splashes reaching the scroll. A high grade official might also keep at hand a jade sceptre (*ju-i*) with flattened head and curved handle of a kind used at court and commonly presented by an emperor to persons of distinction, and in addition a hat-stand with carved domed head mounted on a wooden stem and base set with jade. Elaborately carved vases, figures, and smaller versions of the great jade mountains found in Ch'ing palaces further contributed to a man's standing and enhanced the aesthetic ambience of the higher reaches of Chinese administration. Personal ornaments, including jade buttons of floral design, bracelets, pendants, hair-pins and ear-rings, emphasised still further the link between access to jade and high official status. The introduction of tobacco from America by Portuguese or Spanish merchants during the later Ming period led to the production of snuff bottles made in a variety of materials. A high official would be likely to have one of jade, nephrite for the body and jadeite for the stopper.

Europe

As mentioned earlier in this chapter, jade was prized as a prestige substance among Neolithic communities in temperate Europe at the opposite extremity of Eurasia. Jadeite objects have been recovered over a territory extending from Brittany to the Rhineland as well as over the British Isles as far north as north-east Scotland, far beyond the known sources of the raw material in Piedmont and Switzerland.[40] Apart from a few rings like those recovered from the megalithic passage-grave at Mané-er-Hroek in Brittany,[41] the jade artefacts circulating in Neolithic Europe consisted entirely of flat celts with pointed butts. Their Neolithic context is confirmed by occurrences in the cairn of Cairnholy, south-west Scotland, in the causewayed enclosure at High Peak, Devonshire, and above all on the course of the Sweet timber trackway in the Somerset Levels dated in radiocarbon years to c. 3200 B.C. The absence of any trace of haft or binding under conditions exceptionally favourable for the survival of organic materials and the lack of the slightest sign of wear suggest that the jadeite celt was deposited as an unhafted and unused blade to fulfil a symbolic role.[42] Positive evidence that axes of similar form enjoyed a symbolic status may be cited from one of the megalithic uprights of the Breton passage-grave of Gavrinis. Several pointed-butt celts are depicted by pecking away the surrounding surface of the stone.[43] The question could well be asked why jadeite lost its symbolic role during the Bronze Age. A likely hypothesis is that the material had not been established as a precious substance long enough in Europe to withstand the strong competition of gold.

New Zealand

A third centre of jade-working was that still active in New Zealand as the Maori first came under the direct observation of European explorers, colonists and ethnologists.[44] Here also, as we have seen, the sources of the greenstone were far removed from the densest areas of settlement and the material could be obtained by the inhabitants of the North Island only by more or less hazardous routes. The role of nephrite in Maori society was predominantly symbolic. It served above all to denote the superior status of chiefs against commoners. In this respect adzes were especially significant. New Zealand was amply supplied with basalt, greywacke and argillite, each of which is known to have been used for the adze blades used to shape the timbers required for canoes, houses and defensive works. From a technical point of view the superiority of nephrite over the stones plentifully available for everyday tools would hardly justify the increased cost of ensuring adequate supplies or the enormously greater cost of shaping it to the correct form. For symbolic purposes on the contrary the extravagant cost of nephrite was a powerful recommendation. The ability to dispose of the most costly resources is an essential attribute of power in all stratified societies, and the display of artefacts made from costly materials was one of the most conspicuous ways of demonstrating this power. The symbolic role of greenstone adzes in Maori society is exemplified by the elaborately carved hafts in which they were mounted (plate B) and not least by their designation *toki pou tangata*, meaning the adze which establishes authority. By the same token wood carvings of chiefs show them grasping hafted adzes. (The fact that the Chinese sign for 'father' stems from the glyph of a man and a jade adze suggests that a similar notion prevailed in that country as far back as the Shang dynasty.) Another nephrite attribute of Maori chiefdom was the massive short-handled club or *mere*. In the hands of an experienced warrior such a weapon was capable of smashing human skulls at a blow. Examples made famous by the exploits of their chiefly owners acquired names of their own and passed down the generations as heirlooms. A single example is said to have been sufficient to ransom a chief. When Captain Cook encountered the Maori he found the chiefs adorned with jade insignia. They wore nephrite ear-drops and chest ornaments carved into the form of *heitiki*, grotesque human figures with eyes encircled by shell inlay (Plate F). The

H Tutankhamun's innermost coffin

I Pectoral from the tomb of Tutankhamun

regard in which chiefs held their jade ornaments is shown by the way they stored them along with prized bird plumage in the elaborately carved wooden treasure boxes which rank among the most outstanding products of Maori art.

The New World

Jade occurred in widely separated parts of the Americas.[45] In some of these, for instance California and Wyoming, the material was apparently ignored by the aboriginal inhabitants. Others made use of it for axe or adze blades, among them the Indians of Amazonia, the Haida of Queen Charlotte Island, British Columbia, and the Eskimos of the Kobuk river, Alaska. Hunter-fishers with assured supplies of food and plenty of free time were able to afford the extra labour involved in working such a tough material as jade more easily than working agriculturalists. In the case of the Kobuk Eskimos their object in working jade, which they even carried on while waiting for caribou to cross the river,[46] was to obtain a ready supply of trade goods. When Captain Cook encountered them in 1778 the Eskimos of Bering Strait were using iron knives and wearing glass beads. Martin Sauer records[47] that when he visited the region in 1790 the Eskimos congregated in the present locality of Nome hoping to obtain supplies of precisely these commodities. In exchange they eagerly proffered jade adze blades as well as weapons and articles of clothing.

In Mesoamerica the social situation was quite different. There conditions were more like those prevailing in China. Peoples like the Maya and Aztecs were organized on a hierarchical basis and supported civilizations altogether more developed than those prevailing at the time in North America. Jadeite artefacts were prized above all for their symbolic roles. They served to mark conspicuous waste in the form of offerings and sacrifices and at the same time as objects of conspicuous consumption on the part of individuals highly placed in social hierarchies. Excavations at the Olmec ceremonial centre at La Venta, Tabasco,[48] dating from c. 900–600 B.C., revealed numerous caches containing jadeite artefacts. These were mainly disposed along the main axis of the monument, with a number clustered on platforms raised on either side. One of the richest (fig. 17) included six jadeite axe or adze blades and sixteen human figures, one of granite, two of jadeite and thirteen of serpentine. A feature of the blades is that several showed traces of having been made from river-worn cobbles. Another cache comprised a cruciform setting of thirty-eight polished stone axe or adze blades, four of jadeite and the rest of serpentine. The significance of the blade as a symbol is confirmed by the abundance of jadeite axe-god pendants from the Nicoya zone of Costa Rica, dating from A.D. 500–750.[49] The butt-ends of these objects were shaped to represent folded arms and human heads, often with beaked noses. Personal ornaments, including ear-plugs, necklaces and pendants of jadeite, were also represented among the offerings at La Venta. A foundation deposit beneath a late Pre-Classic (c. 300 B.C.–A.D. 300) platform at Nohmal, Belize,[50] contained jadeite ear-plugs, a tubular bead and a number of pendants in the form of

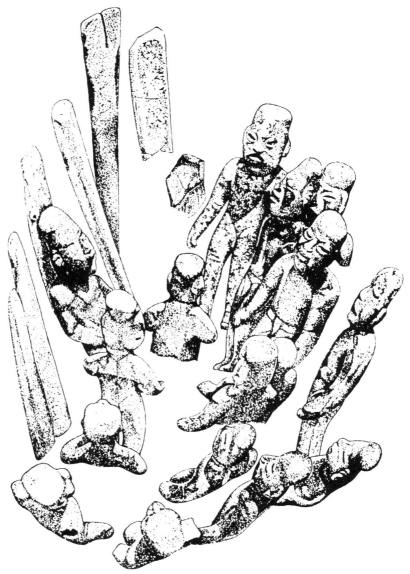

Figure 17 Votive offering at the Olmec ceremonial site of La Venta, Mexico, with stone figurines and jadeite blades arranged upright as a tableau. The figurines vary in height from 160 to 180 mm.

human heads. Jadeite objects also featured among the offerings deposited in the famous sink-hole or cenote at Chichén Itzá in north Yucatan.[51] The role of jadeite as a social marker is best displayed in burial finds.[52] A Late Classic tomb, placed prominently on the main axis of a truncated pyramid west of the West Plaza of the majestic site of Tikal in Yucatan, yielded a splendid chest ornament depicting a head, identified as that of the sun-god Kinich Ahan, wearing a grotesque head-dress, ear-plugs and a necklace resembling those commonly made of jadeite.[53] The writings of Spanish priests dating from the colonial period confirm the high social standing of jadeite among the Indians. Father de Sahagun[54] spoke of this substance as 'precious, esteemed, valuable . . . It is one's lot, the lot of rulers, of the old ones.'

4

Precious metals

Gold

The recognition of gold as a symbol of excellence might almost seem an integral part of human consciousness. In reality it has only become an almost universal standard in the course of history. It owes its unique status to the fact that the people who developed modern science and in many other ways created the modern world community had acknowledged the supremacy of gold since prehistoric times. If it had been the Chinese who engulfed the world in a wide-embracing economic system, we might well have found ourselves using jade rather than gold as a symbol of supreme value.

The primary appeal of gold as of other precious substances was to the senses.[1] People have been and still are attracted by its colour and above all by its lustre. Even so its appearance can be greatly enhanced by the art of the goldsmith. This is largely due to the softness and malleability of the metal. A single Troy ounce can be beaten into a sheet a hundred feet square and only 1/282,000 of an inch thick. Gold is also extremely ductile. An ounce can be drawn into a fine wire some fifty miles long, suitable for making delicate chains and filigree as well as threads for gold lace or tissue.[2] This means that the glories of the metal can be displayed with maximum economy. Moreover the fact that so long as it is relatively pure gold does not tarnish means that its lustre remains undimmed.

Gold is at its most malleable and ductile when it is purest. As a rule it contains varying proportions of base or less precious metals. When the content of silver reaches a half the metal is generally referred to as electrum. For certain purposes gold was alloyed intentionally. Silver might be added to toughen the metal for making jewellery intended for hard wear. In some cases gold might even serve as an alloy itself. Where tin was absent, as in Colombia, it could be used as an alloy for copper to make *tumbaga*, a kind of gold-bronze easier to cast than tin-bronze and capable of reproducing finer detail. Although the addition of gold softens *tumbaga* axes, their working edges could readily be toughened by hammering.[3] It should be emphasized, though, that the prime object of alloying copper with

50

gold and/or silver was to alter its appearance and fit it to serve symbolic rather than technological ends.

Sheet gold was much used in jewellery. Well-known instances include the delicate hair-ornaments worn by one of the ladies buried in the Royal Cemetery at Ur (Plate D), the moon-shaped neck ornaments (lunulae) fashionable during an early phase of the Hiberno-British Bronze Age or, again, the prolific head-ornaments, face-masks and pectorals used in prehistoric Colombia. Thinly beaten gold laid on wooden bases imparted glamour to coffins and large figures and sheet gold could be applied to embellish metal harness fittings or objects of parade. Bowls and drinking vessels could be raised by rotating and hammering gold discs. The technique of hammering meant that sheet metal was liable to become brittle. This led to an early invention of annealing, whereby the metal might be softened by sudden heating and quenching. Other techniques devised by goldsmiths as early as the Royal Cemetery at Ur included pinning, rivetting and soldering.[4]

The varying degrees of purity of native gold involved different melting-points. This led to the early discovery of soldering. The method depended on pouring molten gold with a melting-point lower than that of the pieces to be joined together into the interstices between them: on cooling the separate parts would be found to have bonded together. Soldering was valuable for such purposes as attaching handles to vessels or closing the final links of chains. It was also useful for decorative purposes, attaching the cells for cloisonné work to their background or securing gold droplets to filigree work. The softness of gold made it relatively easy to employ for ornamental purposes. Repoussé work made by hammering up raised areas of sheet gold against a solid block held against the inside and the use of punches and tracers for chasing designs on the outer surface were both being practised by Sumerian smiths at the time of the Ur cemetery. By means of cloisonné work and the complementary champlevé technique, whereby the metal was hollowed out to hold inlays, it was comparatively easy to secure enamels and precious substances as well as artefacts like cameos or intaglios. Alternatively gold lent itself as an inlay for enriching objects made from bronze or silver, such as the famous lion-hunt dagger from Mycenae or the bronze artefacts of Zhou China or the archaistic metal vessels of the Song dynasty.

The high esteem in which gold has been held in most parts of the world, the relative ease with which it can be wrought and not least its capacity for combining with other desirable things have led to its being favoured above all others for jewellery, objects of parade and a variety of insignia of status. The visual splendour and durability of gold which made it an outstanding symbol of excellence were matched by the fact that however widely distributed and keenly sought in nature it has remained rare. Despite the enterprise and labour put into its acquisition it has been estimated that all the gold won between the mid-fifteenth and mid-twentieth centuries could be contained in a cube of fifteen yards.[5]

Gold is much more widely distributed in nature than jade. It has been exploited in every continent and occurs under a variety of geological circumstances. Gold appears either in its parent rock or in placer deposits in which the products of long periods of erosion are concentrated in the alluvium of existing river systems or buried under later sedimentary or igneous rocks. The cost of recovery varies greatly between the two. Alluvial gold, which most commonly occurs as dust or fine flakes, the residue left behind when lighter materials have been removed by the flow of natural waters, can be won by simple sluicing or washing, or even picked up on the surface in the form of nuggets shaped by the compression of fine particles into compact masses by natural forces. Gold obtained by mining pebble conglomerates or veins in a quartz matrix embedded in rock involves the use of greater resources for detaching the ore, bringing it to the surface, pulverizing and purifying it.

It follows that the methods by which gold was obtained differed according to social circumstances. Prehistoric communities and pioneers in new territories concentrated on auriferous alluvial deposits which they could work in small groups or even as individuals with only the most rudimentary equipment. The main drawback of such workings was that by their very nature they were of only limited duration. Greater continuity of supply could be ensured only by mining. Although small scale operations might be undertaken by quite small communities with only simple equipment, intensive mining on a larger scale involved a concentration of disciplined labour, as well as capital resources, of a kind available only to societies organized as states. The differing requirements of working alluvial deposits and mining can be illustrated very well from the situation existing in different parts of Spain during the period of Roman rule. Whereas the recovery of alluvial gold could safely be left to the Iberians of the Guadalquivir, the Douro or the Tagus, the Romans found it necessary to exercise much closer control to extract gold from the mines of Asturias. Conversely the rush to the alluvial deposits of the Klondike in 1896 was carried out by prospectors with little more than shovels and pans. Tapping the immensely productive Main Reef Group of the Witwatersrand[6] on the other hand could be achieved only by mobilizing substantial capital and the most advanced technology.

Despite its widespread occurrence gold has always been scarce in relation to demand. Its scarcity contributed to its status as the most universally accepted precious substance, but this only ensured that unceasing efforts were made to reduce scarcity by increasing supplies. This has only been accentuated in the course of social evolution. The emergence of stratified societies culminating in states increased conspicuous consumption of precious substances. This trend has only been strengthened with the enfranchisement of spending power in modern industrial societies. Again, when civilized states extended their frontiers they frequently took occasion to prospect for and exploit sources of precious substances and most notably of gold. The connection is fully apparent in the case of the Romans. Expansion to north Italy brought into play the gold of the Val d'Aosta and south Piedmont, but it was the Second Punic War (218–201 B.C.) which first increased the supply of gold significantly by taking in the alluvial deposits of the Guadalquivir. By expanding their rule over the rest of Spain the Romans acquired the Douro, the Tagus and the mineral deposits of the Asturias. Acquisition of the Macedonian mines in the mid-second century B.C. brought in further notable gold resources and these were still further enlarged by Trajan's conquest of Dacia early in the second century A.D.

In similar fashion the expansion of European sovereignty overseas, which as much as anything marked the onset of the modern age, was attended by substantial accessions of gold. During the sixteenth century the Portuguese had already been profiting from the gold being produced in West Africa and Japan. Their colonization of Brazil was even more momentous since it opened up a source which during the late eighteenth and nineteenth centuries led the world in the production of gold. The eastern expansion of Muscovite power told a similar story. Peter the Great himself encouraged the quest for gold in the valley of the Oxus. By the middle of the eighteenth century alluvial workings had been opened up in the Ural mountains. Exploitation of the alluvial deposits of the Altai still further east allowed Russia to displace Brazil and for a time to be the world's leading producer of gold.

The predominance of Russia was overtaken during the latter half of the nineteenth century by a succession of gold rushes to more or less remote parts of the world colonized predominantly by the British. The first, that of 1848, was prompted by the recognition of gold particles in a Californian mill-stream. Before the year was out four thousand prospectors had converged on the scene, to be joined in the following year by another hundred thousand, about five times the total white population of the state before gold was discovered there. Many came from within North America itself, but others concentrated from as far afield as South Africa, Australia and China. Between 1851 and 1855 huge quantities of gold were recovered, culminating in 1853 with 200,000 lb. It is an interesting reflection on relative values that when Chinese coolies returned home they took with them greenstone pebbles collected from the Fraser River. One of the Australian diggers on the other hand went home early to apply his newly won knowledge to his home state of New South Wales. Within a week of returning he had struck gold. Before the year 1851 was out a thousand men were at work. The Australian gold rush was on. By 1862 the population of New South Wales had doubled and gold was being recovered in New Zealand, as well as in Queensland, Victoria and West Australia. By the first decade of the twentieth century Australia was yielding 230,000 lb of gold a year. The last of the poor man's gold rushes was that directed in 1896 to the confluence of the Yukon and Klondike rivers, Alaska. The draw of gold attracted men from all parts of the world to a region with an inhospitable climate and provided with only the crudest amenities. By 1897 Dawson City had grown up where no previous settlement existed. In the meantime a new chapter had opened when operations began to recover gold from the Main Reef of the Witwatersrand.[6] The exploitation of this source involved heavy financial commitment and the application of advanced technology. To win a pound of gold the Rand miners had on the average to raise, crush and purify some sixty-seven tons of ore, much of it under extreme temperatures and from great depths.

Although as we have seen earlier the decorative qualities of gold could be explored by direct hammering, the archaeological record shows that goldsmithing of a sophisticated kind in fact developed in communities which practised copper or bronze metallurgy. The Sumerians obtained their gold as well as their copper from the north. Excavations at Tepe Gawra[7] much nearer the highland sources of metals show that close contacts had already been established with the city states of Sumer as far back as the middle of the fourth millennium B.C. By the time of the Royal Cemetery of Ur (c. 2600–2500 B.C.) metallurgy and not least gold- and silversmithing had already reached a stage at which many of the fundamental processes had been mastered.[8]

The ancient Egyptians[9] had ready access to gold, mined as well as alluvial, in the eastern desert, and had already been in contact with south-west Asia during their Predynastic phase. What is certain is that the Egyptian goldsmiths had developed their skills to the point at which they were able to mount amethyst, lapis lazuli and turquoise in the gold bracelets buried with King Zer of the first dynasty in his tomb at Abydos. Nearly one and a half millennia later their skills were tellingly displayed in the decoration of Tutankhamun's golden coffin (Plate H).

The earliest goldsmithing in Europe was also practised in the context of copper and bronze technology. Already during the first half of the second millennium B.C. Minoan smiths were fabricating gold jewellery from metal most probably imported from Egypt. The superbly ornamented gold drinking vessels from

Mycenaean tombs on the Greek mainland (fig. 18) remind us that the artificers of the Aegean world had already acquired almost all the basic skills involved in working precious metals.[10] Goldsmithing was also practised with some skill among the Bronze Age communities of temperate Europe. Even in the peripheral Hiberno-British province they were already displaying a notable capacity for innovation at an early stage of the local Bronze Age.[11] Notable goldsmithing was also practised under the patronage of Celtic chieftains in the La Tène phase of the pre-Roman Iron Age over a territory extending from Switzerland to Ireland.[12] Particularly striking exemplars are the penannular neck ornaments or torcs, the hoops of which were hollow or more often formed of twisted rods or finer wires, having loop terminals decorated by lost-wax casting combined with surface tooling (fig. 19). The Migration Period in Scandinavia witnessed the production of objects made from the great quantities of gold accumulated in the Roman world, much of which moved north when the Empire collapsed. Among the most notable gold objects produced were bracteates[13] or medallions made from thin discs, and massive multi-ring collars decorated with human and animal figures in Germanic style carried out in filigree (figs. 20, 21).[14]

Another early focus of goldsmithing was the Indus basin. Harappan sites have

Figure 18 Mycenaean gold cup from Vapheio, Laconia, Greece. (National Museum, Athens)

55

yielded gold objects[15] mainly in the form of personal trinkets in the context of a metallurgical industry distinct from that of Sumer. In China, on the other hand, where jade was established as the most valued and highly regarded precious substance, gold did not become prominent until Persian influence became important during the Eastern Zhou dynasty and the Han empire,[16] and first really flourished under the Ming and Ching dynasties. Gold appeared during the prehistoric period in Japan[17] but never made a deep impression on the consciousness of the Japanese.

In the New World metallurgy developed comparatively late and then in the main, as we have seen (p. 3), to satisfy symbolic rather than utilitarian needs.[18] The first traces seem to have appeared in Peru at the beginning of the second millennium B.C. By the birth of Christ a relatively complex metallurgy, including

Figure 19 Celtic gold torc from Snettisham, Norfolk. The hoop consists of eight gold wires twisted together, the ends of which have been soldered to hollow loop terminals. These have been decorated with Celtic designs in relief made by the lost wax method but emphasized by hatching. Such objects were probably made by goldsmiths working for chieftains, as with the crown jewels of more advanced polities. Similar torcs were worn by the deities, like those represented on the Gundestrup cauldron (fig. 30), and it is possible that they may have been used to dress wooden idols. Diameter 200 mm. (British Museum)

rich goldsmithery, was being practised in Panama and Costa Rica. Further north metallurgy was apparently absent during the Classic phase of the Maya civilization and was first practised in Mexico around A.D. 800 or 900. The goldsmiths of Colombia relied mainly on hammering in conjunction with annealing, soldering and the use of pins for effecting joins. Occasionally also gold was moulded and for complex forms the *cire-perdue* method was used. For decoration a number of techniques were employed, including repoussé work, granulation and the use of a narrow chisel.

Figure 20 Gold bracteate from Ravlunda, Scania, Sweden. The designs punched on the central roundels of the thin gold discs known as bracteates, several hundred of which have been recovered from Scandinavia, began by reproducing the image of a Roman Emperor and were originally inspired by the medallions presented by the Romans to native leaders who had rendered meritorious service. The image soon came to be reproduced in Teutonic style. Runic inscriptions suggest that bracteates served as amulets. They were thought to bring good luck to the wearer by extending to him something of the protective power of a Roman Emperor. (Lunds Universiteits Historiska Museum)

Figure 21 Gold collar from Alleberg, Västergötland, Sweden. The collar is massive, and its diameter is 203 mm. The portion illustrated in detail depicts a human figure, two animal figures with rich filigree work, and pairs of masks filling gaps between the rings. (Statens Historiska Museum, Stockholm)

Silver

The successful exploitation of silver, indeed the mere winning of the metal on an adequate scale, called for a more sophisticated knowledge of metallurgy. Whereas gold was widely accessible in a form suited for immediate use by the smith, native silver, though doubtless exploited by early man when it occurred in nature, was too scarce in the Old World to serve as the basis for any very substantial industry. Further, it was more difficult to recover. Unlike gold, native silver does not occur in alluvial deposits, but has to be won from veins in mountainous regions or recovered by metallurgical processes from natural alloys or ores. A natural alloy of special importance was composed of gold and silver. This ranged from yellow gold with comparatively low values for silver, through pale yellow electrum with a higher silver content, to white gold which could pass for silver.[19] To obtain anything like pure silver from a natural gold alloy involved a relatively sophisticated process of separation. A source of silver much exploited in early times was lead sulphide, most notably galena, containing varying proportions of silver. The process of extracting the silver involved smelting the ore, purifying the crude lead and recovering silver from the soft metal, notably by means of the cupellation process.[20] The extent to which silver was obtained from galena rather than from natural silver–gold alloys can be judged in part through the presence of lead artefacts and more directly by the purity of silver artefacts. Whereas silver separated from natural alloys is liable to contain perceptible traces of other metals, that obtained by the cupellation process shows up by its high degree of purity.

The archaeological evidence suggests that silver first came into use on a substantial scale during the third millennium B.C. It was not until metallurgy had been carried to a relatively advanced level of sophistication that the production of silver in any quantity became practicable. Deposits of galena are widely distributed but by no means equally accessible to early centres of civilization. Although, for instance, deposits of galena exist between the Nile and the Red Sea and were doubtless drawn upon as a source of the eye-paint known from graves of the Predynastic period, these in fact contained relatively little silver. Down to the end of the Middle Kingdom silver was twice as valuable as gold in ancient Egypt. It was not until the Ptolemaic period that its price fell to that obtaining elsewhere in the ancient world. Mesopotamia depended for its silver on the exceptionally rich galena deposits of Asia Minor, Armenia and Kurdistan.[21] Early Dynastic rulers must have drawn booty or tribute from the silver mountains for centuries before Sargon I sent the expeditions mentioned in documents recovered from Amarna, Assur and Boghazköy.

Until silver objects from archaeological sites have been subjected to more systematically metallurgical analysis one can hardly be sure when and where the metal began to be extracted from galena ores. It is likely that many of the silver vessels from Ur which closely resemble those of gold were in fact made of electrum, a natural silver–gold alloy. The same may well apply to those from the Maikop burial[22] in west Caucasia. On the other hand some, if not all, of the silver from Troy II at the western extremity of Asia Minor was derived from galena ore. Analysis of silver ingots from Troy II has shown them to equal the standard of purity of Roman refined silver.[23] The extensive use of lead during the Cycladic Early Bronze Age points in the same direction for the Aegean. Lead ingots from Phylakopi, Melos and Aghia Irini, Kea, the three votive lead boats from an Early Cycladic grave on Naxos, the use of lead for small figurines and bracelets as well as for rivets used in mending pots, all go to suggest that the silver used in the Aegean during the third millennium B.C.[24] was derived from galena ores. By contrast the rare finds of silver objects from Bronze Age Europe, notably those from Sardinia, north Italy and the el Argar culture in southern Spain, are likely to have been made from local occurrences of native metals or ores. Outstanding silverware was fabricated in the Classical world. The Greeks favoured silver vessels as votive offerings, and the Romans for private ostentation as well as for diplomatic gifts intended to disarm barbarian princes (fig. 22).

Silver played an important role in the Harappan civilization of the Indus basin. It was used for making vessels based on bronze forms as well as for ornaments. Lead ingots argue that the earliest silver used in the Indian subcontinent[25] was derived at least in part from galena ores, notably those abundantly available in Afghanistan. Like gold itself, silver ranked very much below jade in the estimation of the Chinese. It first attained some prominence during the Han dynasty when contacts with the west were intensified. During the Tang dynasty Chinese silversmithing underwent strong influence from Persia, notably in respect of vessels (fig. 23), which in turn affected the shapes of porcelain.[26]

In the New World the working of silver, like that of gold and copper, seems to have begun in Peru where native silver could readily be collected for melting. Although Mexico accounted in recent times for more than a third of the world's production of silver, the extraction and working of the metal began there only late in the first millennium A.D.

Gold and silver owe their nobility to the fact that unlike the base metals they retain their brightness and lustre for an indefinite period in a pure atmosphere. Although harder than gold, silver is still malleable and comparatively easy to work. Since with rare exceptions gold was so much more valuable than silver it was generally preferred for fine jewellery (Plate D). For the same reason silver could more often be afforded for making large vessels, as one can see from products of the Sumerian, Harappan, Aegean and Classical civilizations. The same processes of smithing were employed. The forms of the vessels were raised by hammering and completed by soldering on additional cast or hammered features. Surface decoration was achieved by modelling and lost-wax casting, by repoussé work, the infill of champlevé work by gold or niello or by incising or punching designs. Once a craft like silversmithing was established there was scope for innovation as well as for the transmission of inherited skills. A good example is the way Mycenaean smiths anticipated Bolsover's invention of Sheffield plate in 1743 in the way they capped the rivets used to secure dagger and sword blades to their hilts with silver. Silver caps were bonded with copper rivets by heating the two while held together under pressure.[27] The fact that the efficiency of rivets was in no way improved by capping them with precious metal

Figure 22 Roman silver cup from Hoby, Denmark. One of two silver cups found together with a set of wine-drinking utensils in a grave. The cups, showing bold and sensitive relief, embody the best tradition of silversmithing of the early Roman Empire, and carry the signature of the Greek artist Cheirisophos. On the base they are inscribed with the name of a former owner, Silius, who served as legate of Upper Germany between A.D. 14 and A.D. 21. Costly drinking vessels of this kind were among the diplomatic gifts traditionally bestowed on native chiefs. (Nationalmuseet, Copenhagen)

provides another indication of the limitations of material factors in accounting for technical innovation. The rivets of personal weapons were capped with silver in order to make them better symbols of the superior power and prestige of their owners, not to render them more effective in actual warfare (fig. 24).

Platinum

Although platinum is now the most costly of the precious metals, it is dealt with last because it has not yet attracted the affection or exerted the influence of either gold or silver. A main reason for this is that by comparison its history is so brief. The metal was originally exploited by the Indians of Colombia and Ecuador who recovered it in the form of grains and occasional nuggets from gold-bearing alluvial deposits of rivers draining into the Pacific. Analysis of prehistoric artefacts has shown that the metal worked was of varying degrees of purity. Some things were made from almost pure platinum, others showed small proportions of gold that could easily have arisen from imperfect sorting. Those with a proportion of gold amounting to from a quarter to three quarters must have been intentional alloys. Pure platinum could readily be shaped by hammering, the technique most commonly found on prehistoric nose-rings from Colombia,[28] but

Figure 23 Tang dynasty silver cup, said to come from Lo-yang. A miniature vessel with delicately chased floral design on the exterior, the interior lined with a separate sheet of silver without decoration. The handle was attached by soldering. Height 45 mm. (Karl Kempe Collection)

J The 'Phoenix' portrait of Queen Elizabeth I

K The Chancellor of the University of Cambridge

its high melting point (1775°C) put casting the pure metal beyond the reach of early smiths. The only way to overcome this was to add gold dust, heat the two together until the molten gold bound the platinum granules together, and then by alternately heating and hammering the mixture convert it into a compact mass capable of being forged or cast. Platinum was also used to plate objects made of gold or tumbaga, as in the case of a human mask from the Esmeraldas region of north Ecuador. Similarly the Esmeraldas smiths took advantage of variations in colour to insert platinum eyes into masks of gold or tumbaga.[29]

Platinum first reached Europe in the wake of Spanish colonial enterprise, but it was not until 1741 that it first reached England. There in the course of the next hundred and fifty years rapid progress was made in classifying the properties of the variety of metals included in the group. Up till the outbreak of the First World War the needs of the industrial world could still be met almost entirely from alluvial deposits. It was only as these were run down that recourse was made to mining. At present the world depends for its platinum on refining copper nickel ores from a limited number of mines in Canada and southern Africa. Scarcity, combined with the positive quality of being resistant to wear, has made platinum more valuable than gold. Although it has yet to displace the older established metal in western sentiment, court jewellers have for some time been using platinum as a more appropriate setting for diamonds than gold. This is partly because as a harder metal it makes a more secure setting for valuable stones and partly that being a white metal it reflects rather than competes with the brilliance and lustre of diamonds. Doubtless it was for such reasons that the crown made for Queen Alexandra to display the Koh-i-Noor and the third and fourth Stars of Africa had a platinum rather than a gold frame. More recently the rapid increase in the wealth of the Japanese, their greater consumption of diamonds and the fact that they have never been so attached to gold as the peoples of the west have led to a sharp increase in their use of platinum for jewellery. Today nine out of ten

Figure 24 Mycenaean dagger from Gournia, Crete. The copper rivets have been capped with silver by diffusion, a technique 'discovered' at Sheffield in the mid-eighteenth century and used in the manufacture of Sheffield Plate. Length 270 mm. (University Museum of Archaeology and Anthropology, Cambridge)

Japanese brides choose platinum for their wedding rings.[30] It appears that even now we are witnessing the establishment of a new and uppermost tier in the hierarchy of precious metals. Already platinum nobles have been struck in the Isle of Man to compete with krugerrands as a hedge against inflation.

5

◇◇◇

Precious stones

In the course of the last five thousand years a wide variety of stones other than jade have been highly esteemed for jewellery, insignia, seals and the embellishment of personal weapons. The relative esteem in which stones have been held at different times has rested on a number of factors. Among these colour and texture have been of prime importance, but durability has always been valued in civilized societies concerned with maintaining wealth. Rarity is another component of value and with this often went derivation from distant sources. Exotic substances owed part of their prestige to their mere difference from ones available in the home environment, but even more to the fact that obtaining them from a distance might and in civilized societies normally did reflect the exercise of power.

The attitude adopted to precious substances helps to confirm the force of inertia in history. Once a particular kind of stone was recognized as precious it tended to remain so. This does not mean that orders of esteem remained constant. The discovery of new sources, advances in technique, the opening and closure of markets and not least changes of fashion, no less powerful because often defying rational explanation, all helped to promote change. If old favourites were as a rule long lived, their status tended to be depressed as new substances advanced in esteem. This can be illustrated by comparing the ranking orders recognized at successive periods. It is instructive to begin with to compare the stones listed by the Greek naturalist and philosopher Theophrastus (373/368–c. 287 B.C.), in so far as these can be certainly identified, with those used for the stamp seals of the Aegean world during the second millennium B.C.[1] and the cylinder seals[2] of Mesopotamia as far back as the fourth. Five stones – lapis lazuli, chalcedony, agate, haematite and steatite – are common to Classical Greece and the early civilizations. Three more went back as far as Bronze Age Greece – amethyst, onyx and rock-crystal. In all, eight out of the eleven listed by Theophrastus had experienced long use by gem-cutters. If to carry the comparison further one consults the ranking order of precious stones current in mid-Victorian times set out in Kluge's handbook of 1860,[3] one still finds some degree of continuity but notable changes in ranking order. Thus jade, the material

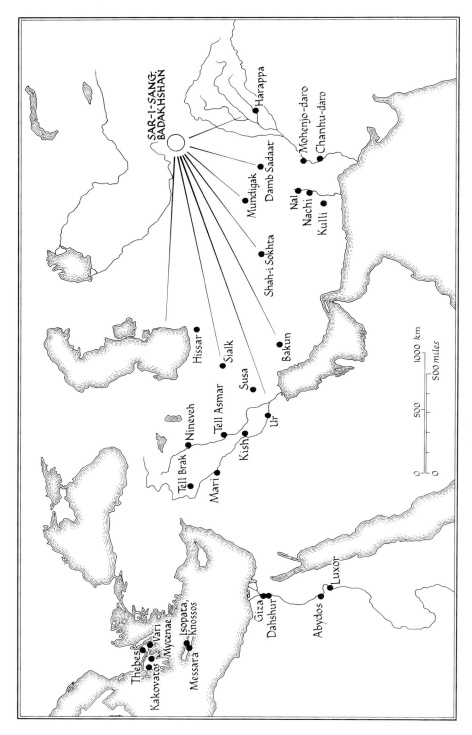

Figure 25 The dispersal of lapis lazuli from Afghanistan.

held most precious of all by the Chinese, was assigned by Kluge to his fifth and lowermost grade on account of its dull colour and lack of transparency. Again, lapis lazuli and different varieties of chalcedony, to which the ancients were strongly attached, fall into his fourth grade along with amethyst and rock-crystal. Of the rest, turquoise ranks in his third and garnet in his second grade. If not a single one featured in Kluge's top grade, this was because of the emphasis laid in recent times on transparency as a mark of excellence. This suggests that it will be useful to review precious stones in two main groups: the opaque to translucent and the transparent.

Opaque to translucent stones

Lapis lazuli

A main reason why lapis lazuli was so highly regarded in antiquity was the brilliance of its colour, blue flecked with gold. The limestone which formed its basis was impregnated with blue minerals, predominantly haüynite, with smaller proportions of ultramarine or lazurite and traces of sodalite. To the Sumerian, Egyptian and Mycenaean peoples lapis lazuli had the added attraction of appearing as an exotic substance. The material occurs in the Old World near Lake Baikal and traces have been noted in the Pamirs, but the only source known to have been exploited in antiquity is that visited by Marco Polo in Badakhshan, north-east Afghanistan (fig. 25).[4] The mines are indeed still operated. Recent reconnaissance has emphasized that this source was not merely remote from the centres which it served, but that it could be operated only under conditions of some rigour. The mines at Sar-i-Sang were situated at over 8,500 feet and could be operated for only a few months in the year. The workings were reached by a precipitous path up which the timber and water needed for fire-setting as a way of detaching the rock had to be carried. Finds at the south Iranian site of Shahr-i Sokhta,[5] as well as much further afield in Egypt, suggest that lapis lazuli was circulated in the form of lumps and worked up at the various centres. Lapis lazuli from Badakhshan is still marketed from Nizhni Novgorod by way of Bukhara. By the Jamdet Nasr period in Mesopotamia lapis lazuli was already reaching Tepe Gawra in the northern part of the country. Around the middle of the third millennium it was featuring strongly among the grave goods deposited in the Early Dynastic Royal Cemetery at Ur (Plate D). It is significant that Queen Pu-abi herself was richly provided with the material, which featured in her head-dress hung around with gold-bound discs of lapis lazuli, as well as forming an important component of her choker and infilling the cloisonné decoration of her fingerring. Gold and silver dress pins had lapis lazuli knobs. Lapis lazuli was also used to inlay the 'standard' from Ur, in reality perhaps the sound-box of a musical instrument, as well as for the horns and upper portions of the fleece of the remarkable cult figure of a goat standing on its hind legs. It was also used to embellish the handle and pommel of a gold dagger buried with its elaborately wrought gold sheath in another of the Early Dynastic tombs at Ur.

The fact that lapis lazuli seems to have appeared first in north Mesopotamia rather than south argues that the Sumerians got their supplies by way of Iran rather than by the Indus and the sea route. The finds from the Harappan sites in the Indus basin,[6] sparse and dating from several centuries later, are more likely to have come independently from the common Afghan source.

Lapis lazuli evidently reached Egypt along with other Asiatic influences during Predynastic times. In the Dynastic period it served primarily to embellish the jewellery worn in life and death by the royal family and household, most notably diadems, necklaces, pectorals and bracelets. Lapis lazuli was only one of several varieties of semi-precious stones used by the Egyptian goldsmith to enrich his products. Strongly coloured ones in particular were employed from Old Kingdom times onwards to infill the cells of gold cloisonné work. Lapis lazuli was valued for its deep cerulean blue, turquoise for its mid-blue, carnelian for blood red and felspar for an opaque green. Other stones used by Egyptian goldsmiths included amethyst, rock-crystal, obsidian, chalcedony, jasper and garnet. Although, apart from lapis lazuli, supplies of all these were available in Egypt and Sinai, Egyptian craftsmen often had recourse to coloured glass to infill cloisonné cells.

As in Mesopotamia and Egypt so in the Aegean the import of lapis lazuli occurred in the context of stratified societies. Aegean civilization, as this appeared during the first half of the second millennium B.C., was sufficiently sure of its own identity to engage in traffic with the older ones of south-west Asia and the Nile Valley. The lapis lazuli scarab from the Second Grave Circle outside the citadel of Mycenae[7] argues that Egypt was one source of supply. Yet there is good iconographic evidence that much of it came from Mesopotamia by way of Syria. Sir Arthur Evans long ago recorded a Sumerian cylinder seal made from this material from a deposit sealed by a Middle Minoan IIIA layer at Knossos. This depicted the man-bull Eabani, seizing an ibex by the horn with one hand and flanked on the other by crossed rampant beasts, as well as a goddess wearing a Sumerian skirt. Evans ascribed this seal to the Syro-Hittite class as he also did a specimen found at Vari, south of Athens on the Greek mainland.[8] More recently a string of no less than thirty cylinder seals made of lapis lazuli has been recovered from a Mycenaean house at Thebes.[9] Some of these were obsolete, but one has been identified as belonging to a Babylonian king, the twentieth of the Kassite dynasty, who reigned from 1381 to 1354 B.C. The suggestion has been made that the string may have been a royal gift.[10] In that case the contemporary seal may have served to validate the emissary. It is worth noting that the seal from Knossos was capped in gold at either end like the one from a Third Dynasty level at Ur. Among the richest finds recovered by Schliemann from Mycenae was a dagger of princely quality from shaft-grave IV with a gold plated hilt inlaid with lapis lazuli insets,[11] recalling that from the much older Early Dynastic cemetery at Ur.

Turquoise

Another stone much favoured in early times was turquoise, an opaque phosphate of aluminium which derives its sky-blue colour from traces of copper. Two derivations have been suggested. Some have attributed it to the old French *torques* in reference to Turkey by way of which the stone reached Europe, and others to the Persian word *piruzeh*.[12] Both are consistent with an origin in Iran where the material is mined from veins in Khorasan.[13] Not surprisingly turquoise has frequently been excavated at sites on the Iranian plateau, notably at Hissar and Yahya.[14] From Iran turquoise was carried as far afield as west Caucasia, where it occurred in the Maikop barrow in the Kuban,[15] and north Mesopotamia, where it was present in the same tomb at Tepe Gawra[16] as lapis lazuli. To the east it reached the Indus Valley, where it has been found at Harappan sites,[17] contemporary once again with lapis lazuli. Turquoise had also reached China by the time of the Shang dynasty, when instead of enriching gold jewellery it was inset on ivory vessels and used to inlay bronze weapons in a kind of mosaic. The ancient Egyptians were also attracted by turquoise, which they could obtain from mines in Sinai. Already in Predynastic times they were using it to make beads, and throughout the Dynastic period it was employed together with lapis lazuli, carnelian and coloured glass to infill cloisonné work on jewellery, notably on the breast ornaments worn by Tutankhamun (Plate I).

Turquoise also met with appreciation in the New World. The Aztecs were using it at the time of the Spanish Conquest in the form of mosaic applied to wooden masks of their gods, combined with shell inlays for eyes and teeth (Plate E). One sign of the importance they attached to this work is that turquoise covered masks featured in the tribute handed to Cortés and transmitted to the emperor Charles V. Some of these passed into the family collection of the Medici and following the dispersal of this the important examples now in the Museum of Mankind passed to the British Museum. Further north in America turquoise was much favoured by the later Pueblo Indians of Arizona and New Mexico, who obtained some at least of their supplies from mines in San Bernadino County, California.

Chalcedony

Several varieties of chalcedony, a silica in crystalline form, translucent and sometimes transparent, waxy to the touch, hard and extremely enduring, were treasured for jewellery, amulets and seal-stones from the earliest civilizations of the Old World down to the present time. One of the most popular has been carnelian, which owes its reddish colour to the presence of iron oxide. This was extensively used by the craftsmen of ancient Egypt as an inlay for furniture and coffins as well as for jewellery and scarabs. Its qualities were also appreciated by the Sumerians – carnelian rings formed part of the queen Pu-abi's head-dress – and by the citizens of the Indus Valley civilization.

One reason why chalcedony has been so highly esteemed is that it lends itself to a variety of decorative treatments. When polished smooth its flattened surfaces, whether flat or convex, offered ideal scope for the engraver. At the same time its hardness meant that it could be used to impress softer substances such as clay or wax without incurring wear. Like the rarer lapis lazuli and some other stones it was particularly suitable for the seal-stones which became increasingly important in transacting the formal business of civilized polities.

A more rare and highly ingenious way of decorating carnelian was invented by the Sumerians. This involved the use of alkaline solutions.[18] Patterns might be drawn on the surface with an alkaline solution, generally soda, and the stone heated, allowing the solution to penetrate and leave a permanent white design on the surface. Alternatively the whole could be whitened by flooding it with alkali and the design etched on, probably with a solution of copper nitrate. Isolated examples have been found as far afield as ancient Egypt and the Indus Valley, another illustration of the extent to which the most precious substances might be distributed to different polities by way of prestige networks.

Banded chalcedony was also popular. Agate, in which the markings were only vaguely defined, was chosen for beads from an early date in Egypt, Sumer and Iran, but it was onyx, in which the bands were more sharply defined and might be strikingly regular, that offered greater scope to the ambitious craftsman. The decorative effect of sectioning onyx is well displayed on a necklace from the Eanna Temple at Uruk dating from the Third Dynasty at Ur. Carefully cut and polished plates of onyx were mounted in gold to form flat components of the necklace and one was engraved with the name of the priestess Abbabashti of Shesin.[19] Sumerian gem-cutters also took advantage to a limited degree of another way of exploiting the banded structure of onyx. This is exemplified by a circular dome-shaped bead from Ur shaped in such a way as to expose one of the pale layers as a ring close to the perimeter.[20] Hellenistic craftsmen concentrated much more intensively on exposing the surface of prominent white bands and carving one or more in relief using the intermediate darker zones to throw their images into stronger relief. In the cameo the Greek craftsman and his Roman colleague expressed in miniature the ideal forms which in marble sculpture serve for many as the material embodiment of the Graeco-Roman genius. The numbers in which cameos of the Classical period have come down to us is one testimony to the regard they have inspired in men of ensuing ages. The two-layered cameo of Augustus[21] which has ended up in the British Museum bears tangible evidence of regard, even of awe, in the delicate circlet of gems, including a miniature cameo, affixed to it while in the hands of a medieval owner (fig. 26). Another sign of attachment is the revival of the art of cameo-cutting, beginning with the commissions bestowed by popes, prelates and connoisseurs at the time of the Renais-

Figure 26 Classical sardonyx cameo of Augustus with medieval circlet. (British Museum)

sance, persisting through the period of bourgeois dominance and continuing into the present age of enfranchisement even to the point at which resort has had to be made to substitute materials like shell or paste to satisfy a greatly enlarged market.

Opals

The opal,[22] the most recent addition to the select company of precious stones, is a hardened gel deposited from a silica-rich solution in the fissures and seams of different kinds of rock. The opalescence which claims attention arises when rays of light strike thin films having an optical density differing from that of the main mass. This is most apparent in the precious opal which is capable of displaying a veritable kaleidoscope of colours. When the Spanish conquerors entered Mexico they found that a special variety, the fire opal, having a clear orange or yellow body, was held in high esteem by the Aztecs. Opals occur in several parts of North America but as a rule these are only of common, non-precious character. The most precious opals, including black opals the rainbow colours of which are set off against a sombre background, are those from the opal fields of Australia opened as lately as 1872 but not seriously exploited until the twentieth century.

Before the New World was discovered and Australian resources were opened up the only source of opals known was situated in the Libanka and Simonka mountains north of Kosice in eastern Slovakia. Opals were totally unknown to the early civilizations of the Old World and only came to the knowledge of the Mediterranean peoples as an outcome of Roman military expansion into the Balkans. Apart from fire opals, which can stand being faceted, opals can be prepared for jewellery only by the age-old techniques of grinding and smoothing, owing to their tendency to crack.

Transparent stones

We owe the next great advance in the appreciation of precious stones to the Hellenistic and Roman heirs of Alexander's conquests which served to promote new dynasties and enhance trade between the Mediterranean and India. To this we owe the diffusion not only of pearls but of transparent precious stones,[23] notably red garnets, green emeralds, blue sapphires and, crowning all, diamonds. By contrast with diamonds, which were too hard to be ground smooth and could only be mounted as natural crystals until medieval lapidaries had learned how to cut them, coloured transparent stones continued to be shaped by the time honoured methods originally devised for opaque stones. It was only when lapidaries had discovered how to release the fire of diamonds by cutting them at angles devised to admit and reflect the maximum amount of light that the coloured varieties of transparent stones were subjected to the same treatment. Meanwhile Hellenistic, Roman and early Medieval jewellers could only treat transparent coloured stones by rubbing them smooth.

Garnets

The first coloured transparent stones to be used for jewellery were garnets. These form a group of double silicates made up of varying pairs of minerals grading into each other without necessarily showing sharp boundaries. By no means all were sufficiently attractive for jewellery and in practice those used for this purpose were mainly stones of red colour, notably blood-red pyropes and dark red almandines. The first people to appreciate them were the ancient Egyptians, who were able to draw on supplies from Sinai. The recovery of a gold diadem set with garnets as well as jadeite, malachite and turquoise from a Late Predynastic grave at Abydos[24] shows how early they had begun to do so. Hellenistic and Roman jewellers continued to rely heavily on garnets,[25] though whether they supplemented Egyptian stones with ones from more remote sources in India or Sri Lanka remains unsure. The Migration period which intervened between the decline of Rome and the emergence of European nations witnessed a marked change in the way garnets were used to embellish gold work. Garnets continued to be rubbed smooth and set *en cabochon*, but advantage was taken of the way the stone could be split into sheets to use it extensively for infilling gold cloisonné work,[26] a technique invented by Sumerian goldsmiths and brilliantly employed, using a variety of infills, in New Kingdom Egypt. The effect of using garnet as an inlay can be seen with particular brilliance in the jewellery purse components, shield ornaments and sword-fittings from the royal ship-burial at Sutton Hoo (fig. 27).[27] These incorporated over four thousand pieces of garnet individually cut so as to fit precisely into the cloisons for which they were designed. During the eighteenth and nineteenth centuries garnet jewellery, much of it then using cut stones, enjoyed a notable vogue among the rising middle classes requiring red stones but unable to afford rubies until these became available in good synthetic form during the present century.

Emeralds

The best known transparent varieties of the mineral species beryl include the velvety green emerald and the pale blue to greenish blue aquamarine. Neither is to be found in alluvial sands or gravels and both have to be mined from the parent rock. Although they share the same chemical composition and crystal form, emerald is by far the more valuable. This arises mainly from the fact that whereas aquamarines may occur in large flawless masses in the rock, flawless emeralds are extremely rare. So much is this the case that the best emeralds exceed diamonds of comparable size in value. The richest sources exploited today are the mines of Colombia, where the Spanish conquerors encountered them in 1537. Significant finds have since been made in the Urals and South Africa. The emeralds used in the ancient world were all mined in Upper Egypt[28] and it is likely that these were first exported on a significant scale to satisfy the Roman appetite for coloured transparent stones.[29] Emeralds were sometimes used in their natural form as hexagonal crystals, but their comparative softness made it a simple matter to rub them smooth and set them *en cabochon*. When emeralds began to be cut, lapidaries found that their qualities were likely to be displayed to best advantage by step-cutting, and this method is still widely used for this stone, though brilliant cutting is sometimes applied.

Figure 27 Detail of purse lid from the early seventh-century ship burial at Sutton Hoo, Suffolk. The cloisonné work is inlaid with garnets and millefiori glass. (British Museum)

Transparent stones

Corundum (sapphire and ruby)

As a mineral, corundum has proved its value to man partly as an abrasive, which allowed it to play a key role in the shaping of jade, and partly because it has contributed two of the most keenly sought after transparent coloured gems, sapphire and ruby. The most important Old World sources of corundum are located in Thailand, Burma, Sri Lanka, Kashmir and Afghanistan, which shared the characteristic that they could most readily be tapped by the India trade to the west. Sapphires owed their attraction to their blue colour, recalling that of the long-treasured lapis lazuli, and were eagerly appropriated by Hellenistic and Roman jewellers.

When rubies began to play their part remains uncertain. Supplies of red stones were already available in the form of garnets, and it is known that these were much used by Hellenistic and Roman jewellers. At present rubies vary greatly in esteem according to their colour and source. Those from Thailand are commonly too dark and those from Sri Lanka too pale. The most highly prized rubies were pigeon blood stones from upper Burma, which are commonly more valuable, and in the case of the rare large stones very much more valuable, than diamonds of comparable size. Although rare, rubies were certainly used in Europe during medieval times. Rubies, along with emeralds, oriental sapphires and pearls were used to enrich the brooches and gold crowns owned by Edward I's wife, Eleanor of Castile, at the time of her death in 1290.[30]

Diamonds

Although the quality of diamonds[31] which most immediately strikes the modern imagination is their capacity to command attention as symbols of majesty or affluence by reason of their fire and lustre, it was their quality as the hardest substance on earth that first engaged attention. Until men had learned how to cut diamonds, they were not particularly attractive crystals. When they first obtained them from India the Romans mounted them as finger-rings[32] less for their appearance than as symbols of hardness and strength. The Chinese took a more practical line by using them to tip drills for perforating jade, anticipating their use in modern industry to fulfil a number of different roles from drilling steel components to drawing filaments.[33]

Until the South African mines were opened up during the last decade of the nineteenth century, diamonds were obtained exclusively from alluvial deposits, often those which also produced gold. India, mainly the eastern zone of the Deccan, held a monopoly in the supply of diamonds until stones were discovered in Brazil in 1725 and brought home to Portugal three years later. Only half a generation later the Dutch had found them in Borneo and more were revealed in 1829 in the Urals as part of the Russian drive to open up the mineral resources of the interior. The mid-century gold rush to Australia added further to the supply of diamonds, but it was the development of the South African mines that for the first time brought a dramatic increase in the volume of production.

If diamonds are the hardest of minerals they are also among the most cleavable. From the thirteenth century European lapidaries conducted active experiments in cleaving diamond crystals and in learning how to shape them in ways best calculated to admit and reflect the maximum light (fig. 28). By 1412 the Duke of Burgundy possessed table-cut stones with flat polished faces bevelled by oblique facets at the sides meeting to form four points. As the lapidaries of the Low Countries gained in recognition and were able to apply more specialized skills they embarked on further innovations. By around 1520 they had developed the rose-cut or rosette by careful faceting of the convex face of a flat-based stone. Similar cuts were used by Indian lapidaries, at least as early as the seventeenth century. Since Europeans could obtain their diamonds only from India some technical interchange must have occurred. Yet if the lapidaries of Europe and India shared the idea of faceting the convex face of flat-bottomed stones, they had to do their work subject to different expectations. Whereas the Europeans aimed first and foremost at obtaining more brilliant jewels, the Indians were evidently constrained by the requirement to conserve the maximum weight of

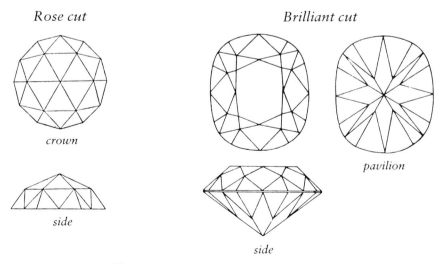

Figure 28 Rose and brilliant diamond cuts.

diamond and therefore aimed to cut more numerous and shallower facets to form a more dome-like gem. When stones like the Koh-i-Noor came into British ownership they would be re-cut to the rosette or brilliant pattern even at the cost of a considerable reduction in weight. The development of the brilliant cut supposedly under the patronage of Cardinal Mazarin during the seventeenth century intensified the effect of diamonds, but entailed a greater loss. The Brazilian Star of the South, originally of 261.88 carats, suffered a reduction of more than half to emerge as a brilliant of 128.8 carats.

A final attribute of diamonds which still further enhanced their value was their durability. This ensured that individual diamonds of any size were liable to have long and sometimes dramatic histories, reflecting at times the rise and fall of empires or the end of dynasties. The added fact that diamonds could be readily cleaved means on the other hand that the histories of individual stones are sometimes difficult to trace.

Enamel

Although not strictly speaking within the scope of this book, since it is an entirely artificial substance, enamel needs to be mentioned if only because it served as an alternative to natural stones in enriching jewellery and other symbolic objects. Indeed enamel was frequently used in combination with precious stones to enhance the same item. Enamels of permanent gem-like quality can be made by adding the appropriate metals to powdered glass fused to a metal base. The medium in its various colours could be used for cloisonné work or painted thinly on metal surfaces. In either case it could be used to portray symbols, figures or scenes. The problem of tracing the invention of enamel is made more difficult by failing to distinguish it more certainly from glass. Careful examination of ancient Egyptian cloisonné work has shown that what was once claimed as enamel was in reality coloured glass skilfully cut to fit the cloisons.[34] Again, analysis of the coloured material used by Celtic Iron Age craftsmen to decorate bronze artefacts is not, as was once supposed, enamel but coloured glass imported from the Mediterranean.[35] On present evidence it appears that Classical Greek craftsmen were the first to employ enamel,[36] but that it was the Byzantine Greeks who carried the craft to a new level of accomplishment[37] in the service of the imperial court and the rituals of eastern Christianity. Crowns, insignia, crosses, reliquaries and the bindings of liturgical books were dignified and enriched by enamelled iconography studded with pearls and cabochon gems. The taste for enamel spread to western Europe by way of church and state. Both the German Imperial crown completed by Conrad II (1024–39)[38] and the Holy Crown of

Hungary (see fig. 35), thought to have been assembled early in the twelfth century[39] incorporate enamel plaques of Byzantine origin. Since enamelling was taken over by the west to produce jewels like that which tradition claims was presented to his foundation, New College, Oxford,[40] by William of Wykeham early in the fourteenth century (fig. 29), enamel has held its own in the embellishment of symbols of high achievement. When civic and other dignatories assume their badges of office or members of orders of chivalry their decorations, they might well enhance their satisfaction by reflecting on their debt to the only seemingly remote hierarchs of Byzantium.

Pearls

Of the two main sources of pearls[41] those from marine molluscs have always been preferred to those from freshwater. The supreme quality of marine pearls which caused the Romans to give them priority among all other materials for jewellery is their lustre or 'orient', counterpart of the 'fire' of diamonds. Size, colour and perfection of form help to influence the value attached to pearls, but it was their 'orient' which made them outstandingly attractive to men of many civilizations.

Pearl-bearing oysters can readily be gathered by divers without recourse to elaborate equipment. They rarely live in temperatures of less than 25°C (75°F) and the banks on which they grow are mainly between three and five, and only occasionally more than five, fathoms deep. The earliest history of the use of pearls is difficult to establish. By comparison with gold, jade or precious stones pearls are much less durable. Those buried with a daughter of the Roman general Stilicho around A.D. 400 are said to have fallen instantly to dust when her tomb was opened in 1544.[42] To some extent mother-of-pearl, which lasts much better, can provide a clue to the exploitation of oyster beds. Wall paintings can sometimes help. Literary sources may be even more valuable, especially those relating to people like the Romans who held pearls in exceptionally high regard.

Among the first people known to have used pearls for jewellery were the ancient Egyptians,[43] who wore them as pendants to earrings and threaded onto necklaces alongside cowries, coral, scarabs and precious stones certainly as early as the middle of the second millennium B.C. Tomb paintings depict them wearing pearls on their clothing and chest-ornaments of mother-of-pearl suggest that they were already exploiting the Red Sea fisheries, later mentioned by Strabo and other Classical writers, by the twelfth dynasty. Yet pearls did not feature prominently in Egyptian jewellery until the country was conquered by the Persians during the fifth century B.C. and direct links were established with the fisheries

Figure 29 The 'Founder's Jewel', New College, Oxford. Although commonly known as the Founder's Jewel, it is doubtful whether it was in fact bequeathed to the College by William of Wykeham. It tallies exactly with the description in the *Liber Albus* of a jewel given to the College by Thomas Hyll, who died in 1468. The jewel is formed by a Lombardic letter M framing on one side the Virgin and on the other the Angel of the Annunciation. It is embellished by pearls and by emeralds and rubies *en cabochon*, and the angel's wings and the lily are picked out by enamel.

of the Persian Gulf. Similarly it was Alexander's conquests which brought pearls to the attention of the Classical world. The Greeks were fond of using them for mounting on earrings, but it was the Romans who developed a passionate interest in pearls. Roman ladies liked to sleep with their pearls and Caligula thought them worthy to decorate his favourite horse.[44]

The Roman appetite for pearls was such that they supplemented imported marine pearls by seeking out those from the freshwater mussels which grew in the rivers of their temperate provinces, despite their relative dullness and lack of lustre. It was even suggested by Suetonius that reports of freshwater pearls in British rivers helped to persuade Julius Caesar to undertake his invasion. Caesar's interest is confirmed by his action in dedicating a breast-plate studded with British river pearls to Venus Genetrix on his return to Rome. (It is interesting to note that freshwater pearls were known to the Chinese by the beginning of the first millennium B.C.)

It is still not known exactly when the oyster beds on the Arabian shore of the Gulf began to be worked. A hint that the fishery may have been active as early as 4200B.C. is offered by the serpent-like ornament of mother-of-pearl recovered from level V C at the site of Yahya in south Iran.[45] The high quality of the Gulf pearls and the fact that the fisheries were controlled by Arabs ensured that from the market at Bandar e Lengeh on the Iranian coast not far west of the Gulf of Hormuz they entered into the trade network of the Indian Ocean and adjacent areas. It was probably as a result of Arab contacts that pearl fisheries were opened up on the coast of East Africa. Other important fisheries were started on the west side of the Gulf of Bengal, notably on the Coromandel coast and in the Gulf of Mannar north-west of Sri Lanka. The produce of the Bengal, as of the East African, fisheries was handled by the Madras market which served to satisfy the Indian demand for pearls for jewellery and decorating clothing. Particularly fine white pearls were produced by the fisheries of Celebes and North Australia. The fisheries of the Suli islands between North Borneo and the Philippines, though not particularly rich in pearls, produced fine mother-of-pearl, a material keenly sought by the Chinese who used it for making inlays. The Pacific islanders, who had their own sources, used mother-of-pearl notably for fish-hooks and ornaments. When the Spaniards reached Middle America they found the native peoples using pearls and mother-of-pearl obtained from fisheries in the Caribbean, the Gulf of Mexico and the Pacific coasts of south California and Mexico.[46]

Although the status of pearls among other precious substances varied in different parts of the world, their appeal was exceptionally widespread. It is significant that the Romans gave them first place, whereas the Chinese valued them above gold, inferior only to jade. At the time of the Han dynasty, when jade was still the only material fit to accompany the emperor, it was accompanied in the case of feudal lords and officials of grades 1–3 by pearls and in that of officials of grade 4 by gold.[47] One indication of the status of pearls in Christendom is their use in iconography as symbols of regeneration, and the way they have long been used to enrich the crowns of sovereigns from the sacred crown of Hungary to the mitre crown of Catherine II of Russia and in our own day the State Crown of Queen Elizabeth II (Frontispiece; figs. 35 and 37).

6

The symbolic roles of precious substances

The symbolic roles of precious substances in human affairs are many and various and they can be viewed from several directions. What is common to all is that as symbols of excellence precious substances embody and display values and it is by entertaining and expressing values that men declare their identity as human beings rather than as mere primates. Precious metals and stones and objects fabricated from them serve as conspicuous trophies of successful emulation, as gauges of affection and assurances of personal worth and well-being. At the same time they define the place of the individual in society, whether in the family or in wider spheres culminating in polities. Equally they sustain as well as designate those who discharge public functions. The confidence and effectiveness of individuals in acting out the roles of magistrate, priest or sovereign are enhanced by the esteem in which the precious substances embodied in the symbols of their offices are held in society at large (Plate K).

Aesthetic appearance, rarity and durability may have been the physical attributes which led particular substances to be categorized as precious, but it was the fact of being recognized as such which made them effective as symbols. The hierarchies of esteem in which they were held differed widely from one culture to another until all were engulfed in the pecuniary measure of values common to the world market. Yet the transmission of precious substances in the form of jewellery or other objects of display has at all times and most notably during the last five millennia served the same purpose the world over, that of signalling and enhancing status. To take a homely example, the mug and napkin ring of silver given at christening in western Christendom symbolize the essential and prime tasks of the infant to eat, drink and start on the way to adulthood. The gold watch presented at coming of age marks the assumption of an adult role. The diamond ring presented by two out of three Anglo-Saxon swains to their fiancées[1] declares the enduring nature of their mutual love, just as the fact that the ring bestowed at their wedding, traditionally made of untarnishable gold, marks the entry into a permanent relationship. As couples pass decennial milestones in their conjugal lives the occasions are marked, if not always in material form, by recognition in terms of the hierarchy of precious substances commonly accepted in western society – silver, ruby, gold and diamond. In such a manner individuals, like institutions or sovereigns, are fortified in their passage through time by jubilees symbolized by precious substances held in progressively higher esteem.

As well as marking stages in the life cycles of people, gifts of precious substances embodied in jewellery and other personal things also serve to enhance family solidarity on fixed occasions in the religious and social calendar. A recent analysis of the American market in diamond jewellery has shown that nearly half is given for Christmas as compared with an eighth each to mark birthdays and wedding anniversaries. The role of precious substances in cementing family relationships can be illustrated particularly clearly from the sentimental phase of European taste prevailing in the mid-nineteenth century. The bracelet made in Paris around 1845 for Princess Eugenie of Hohenzollern-Hechingen[2] was formed of enamelled gold links corresponding with the immediate family circle. Each link was set with a large stone according alphabetically with the lady's husband and siblings – malachite for Maximilian, aquamarine for Amelie, amethyst for Augustus, emerald for Eugenie, chrysoprase for Constantine, jacinth for Josephine, topaz for Teolinde and amethyst for Auguste Amelie. As a final touch each was accompanied by a lock of the particular loved one's hair.

The excellence attributed to precious substances which made them so useful both as symbols of successful emulation and as expressions of love and regard extended to other matters of keen interest to individuals.[3] Particular substances were commonly held to possess magical and even medical potential. So long as these suppositions were taken seriously, they were not only reassuring, but frequently effective. If anyone believed that by wearing, bestowing or even consuming particular substances it was possible to avert poison and other evils or positively to advance a cause, it was comforting enough to ensure the continuity of such beliefs. At a popular level this was done through the transmission of beliefs and practices through folk-lore. For the literate community such, along with extracts from earlier written sources, might be incorporated in the lapidaries which served to accumulate and transmit beliefs about different substances. In his *Materia Medica* the Greek Dioscorides listed some two hundred kinds of stone, including oxides, suggesting that friable ones be reduced to powder for taking internally and hard ones worn as amulets. He recommended coral for skin troubles, sore eyes and bites,[4] to choose a single example. For many Greeks the dangers of ship-wreck and storms were common sources of anxiety. A nautical lapidary accordingly recommended a variety of precious substances to ward off such dangers. Amulets of carbuncle and chalcedony were approved for children's wear at sea, and mariners were advised to bind coral to mast-heads with strips of seal-skin as a precaution against ship-wreck and tempest.[5] In his compendium on natural history Pliny the elder incorporated a vast store of popular beliefs about the magico-medical application of precious substances for the relief of physical ills. He noted the external application of gold in cases of fistulas, piles, ringworm and ulcers and the practice of burning jet to keep serpents at bay and dispel hysteria. He also recorded the belief among women of the Po Valley that amber necklaces were good for curing goitre.[6] The therapeutic properties of amber were widely recognized in the classical world. Pliny himself advised amulets of amber against delirium and recommended powdered amber for the stomach. Hippocrates canvassed oil obtained by the destructive distillation of amber for asthma, as a stimulant and in the form of a liniment for the chest.[7] Pliny's standing as author of the *Historia Naturalis* was such that not a few of the beliefs which he set down without personal commitment have continued to influence popular superstitions for nearly two millennia. As recently as 1922 Dr Joan Evans noted that 'Amber necklaces may still be sold in the chemists' shops of Mayfair as a cure for croup, asthma and whooping cough.'[8]

Although continuing to draw upon Classical as well as earlier Christian sources, the lapidary compiled in Latin hexameters by Marbode, bishop of Rennes, between 1067 and 1081 sought to lay a greater emphasis on moral themes. Among the many precious stones endowed with special powers the bishop singled out chalcedony for its ability to promote victory, topaz for healing, carnelian for restraining anger and amethyst for preventing drunkenness. The most versatile stone in his long catalogue was sapphire, which he held to be good for protecting the limbs from injury and the wearer from fraud, as well as for overcoming envy, averting terror, liberating from imprisonment, purifying the eyes, cooling the body and not least for the convenient property of making the wearer beloved of god as well as of men.[9] Such claims, which would be dismissed today as unscientific, were taken seriously in the past even by the great, who were no less willing than the humble to accept as true what brought them comfort. When Ivan the Terrible felt himself drawing near to death, we are told, he would have himself carried up to his treasury. Calling for his most precious stones he would draw confidence from extolling the virtues distinctive of each kind. In particular, as the traveller Sir Jerome Horsey reported,[10] he would seek out his 'staff roiall, a unicorn's horn garnished with verie fare diamondes, rubies, saphiers, emeralds and other precious stones that are rich in valleu . . .'.

Although the seventeenth century saw the beginning of the end of magic in western society, some credence was still given to the supposed therapeutic properties of certain precious substances. There are even signs that in some cases this was reinforced by reports which reached Europe as an outcome of the voyages of discovery that to a significant degree marked the onset of modern history. The use of jadeite amulets for renal complaints in seventeenth-century England stemmed ultimately from the Aztecs, whose use of the material for this purpose aroused the interest of their Spanish conquerors and, as we have seen, gave rise to the word jade. It is likely also that the practice of swallowing a paste made of powdered pearls indulged in by Francis Bacon[11] among others was derived from India, where it was established in Hindu usage.[12] It is important to remember that, even when the use of precious substances for magical or medicinal purposes was rejected by educated people, it survived much longer in popular superstitions. Decisions in regard to the purchase of jewellery are still not determined only by aesthetics, emulation, tax avoidance or continuing inflation. Whether they realize it or not many consumers are motivated by superstition. It is doubtful how many of those who bestow gifts of coral on infants appreciate that their forebears did so to protect them from bewitchment.[13]

The concern of the group for individual members found expression through precious substances in death as well as in life. A similar ambivalence prevailed in the provision of grave goods as in the bestowal of presents. The burial of jewellery and other appurtenances incorporating precious substances symbolized continuance for the dead person of the degree of esteem enjoyed in life. At the same time it afforded a public opportunity for emulation in what Veblen would have termed conspicuous waste. Families, clans, states and other collectivities could vie with each other in the riches they were able to consign to the graves of departed members. The fine things recovered from the royal burials at Ur, from the tomb of Tutankhamun, the dynastic cemeteries of China or the ship-burial of Sutton Hoo were measures not merely of piety but of emulation at the highest level. So intense was this that the Chinese hierarchy, anxious to preserve the established order, found it necessary to restrain competitive display of grave goods by sumptuary regulation.

When the Spanish pioneers penetrated the interior of Colombia during the early sixteenth century they were impressed by what they took to be the passion of the Indians for the well-being of their dead. According to Pedro de Cieza's account of 1554:[14]

The Indians were buried with as much wealth as possible, and so they strove with the utmost diligence throughout their lives to acquire and amass all the gold they could, which they took from their own land and were buried with it . . .

If Pedro failed to appreciate that the Indians were motivated by emulation as well as by piety in accumulating gold and then sacrificing it in burials, he was well aware of what the practice meant for his fellow countrymen. In the course of three and a half years they were able to recover over fourteen thousand ounces of gold, three quarters of it fine, from the single cemetery of Sinú.[15] By the same token archaeologists have much to gain from this propensity. Whereas the excavation of settlements can throw light on the useful activities of ancient societies, it is from burials not previously looted by treasure-seekers that archaeologists have recovered artefacts shaped by the finest craftsmen from the most precious substances, veritable trophies of the forces of emulation responsible for the rise of human civilizations.

Although precious substances have done much to dignify and enrich the lives of individuals, their prime function has always been social. Even gifts made to

Figure 30 Silver cauldron from Gundestrup, Denmark. This magnificent object was found by a peat-digger in dismantled condition, the embossed plates being stacked in the bottom, a sign that it was not intended to be recovered but was deposited as an offering. Its size, elaborate smithing and the fact that it was made of silver meant that the offering must have been exceedingly valuable. The vessel was probably made in the East Celtic area during the second or first century B.C. The representations give some insight into the deities imagined at the time as well as into some of the rituals involved in their worship. Goddesses as well as gods were envisaged and it is notable that all the former and some of the latter are shown wearing torcs or penannular neck-ornaments with expanded terminals like that from Snettisham (fig. 19). A scene on the inner face of a plate in the upper middle of the aspect illustrated shows a procession of spearmen on foot and mounted warriors accompanied by three trumpeters assisting at what appears to be the sacrifice of a prisoner. (Nationalmuseet, Copenhagen)

individuals are as a rule intended to strengthen family links. In the public domain they are more overtly directed to promoting social ends. The notions of excellence they embody are used to fortify and legitimize the sacred and profane dimensions of authority while at the same time denoting official executants by means of appropriate insignia.

The earliest and most widespread manner in which precious substances were used to influence the unseen powers behind the visible world was by means of sacrifices and votive offerings. The evidence from prehistoric times can seldom be interpreted with any certainty. A hole in the ground may serve as a safe-deposit quite as well as a receptacle for offerings. When precious things like the great silver cauldron embossed with gods, goddesses and cult scenes from Gundestrup in north Jutland,[16] are recovered by peat-diggers in bogs (fig. 30), the votive hypothesis might seem more plausible. It is even more so when treasures were cast into deep water. The carved jades recovered from a well at Chichen Itza in Yucatan can fairly be accepted as offerings.[17] The same can be said of the gold treasures thrown into the lagoon of Guatavita, Colombia, in the course of ceremonies to mark the appointment of successive rulers.[18]

The evidence from shrines, temples and churches erected to meet the needs of literate societies is even more decisive. Written records confirm the evidence of archaeology that foundation offerings of precious substances were made during the construction of early temples in Mesopotamia.[19] Under each of the two surviving corners of the Early Dynastic temples within the oval enclosure at Khafajah the excavator found deposits including gold, copper, lapis lazuli and slate, together in one instance with a block of carnelian. The kings of Assyria[20] were accustomed to offer foundation sacrifices of precious substances. A stone tablet of Samsi-Adad I (1813–1781 B.C.) boasted that he had raised the 'walls of the temple upon silver, gold, lapis lazuli and carnelian'. Over a thousand years later Sennacherib (704–681 B.C.) is recorded as having deposited identical materials along with chalcedony, alabaster and malachite in the foundations of a structure at Assur.

Once constructed, sanctuaries and temples of whatever cult continued to receive votive offerings, and the richer and more prestigious they became the more emulation they aroused and the greater the inflow of precious substances. In the world of Classical Greece, for instance, silver plate began to be dedicated in sanctuaries during the Archaic period.[21] Success in the Persian wars and the establishment of control over the Thracian mines brought further enrichment to the treasuries. An inventory of 418–417 B.C. recorded a hundred and sixty-three sacrificial dishes of silver in the Treasury of Athena. By the second century the temple of Apollo at Delos had accumulated as many as sixteen hundred. Despite the temptation to accumulate treasure it was recognized that one way of stimulating gifts was to spend a sizable proportion on the further embellishment of buildings and the enrichment of furnishings and vestments. For this reason temples and cathedrals frequently maintained artificers schooled in the use of precious substances. As early as the Third dynasty of Ur officials were designated to supervise the production of gold vessels and jewellery for use in the temple.[22] A similar situation must have existed in Peru at the time of the Spanish Conquest. The temple of the Sun at Cuzco appeared a veritable gold-mine to the impatient conquerors. In Prescott's vivid account:[23]

Gold, in the figurative language of the people, was 'the tears wept by the Sun', and every part of the interior of the temple glowed with burnished plates and studs of the precious metal. The cornices, which surrounded the walls of the sanctuary, were of the same costly material: and a broad belt or frieze of gold let into the stone-work encompassed the whole exterior of the edifice.

The western wall of the temple was dominated by a representation of the god engraved on a 'plate of gold of enormous dimensions, thickly powdered with emeralds', the most valuable precious stones of the region. In addition all the plate, ornaments and utensils of the temple were made of gold, including even the implements used in tilling the temple gardens.

Precious substances have enriched religious iconography over a wide range of faiths from the idols of marginal societies to the divinities of religions disposing of priestly hierarchies and sumptuous architectual settings. In the ambience of Christianity ivory was particularly sought after for representations of the Virgin Mary and of Christ crucified. Gold was used not only for covering images but for such auxiliary but revealing purposes as providing members of the Holy Family and Saints with haloes, encircling the head in paint, mosaic or sheet metal. By the same token contracts drawn up for Italian fresco painters during the Renaissance commonly stipulated the use of ultramarine containing powdered lapis lazuli for the Virgin's cloak. Christianity also excelled in the use of precious substances for enriching the furnishings, vestments and accoutrements of religious ritual. The most valued possessions of medieval shrines were the relics which attracted the veneration and offerings of pilgrims. One way of showing them respect was to enclose them in reliquaries and enrich these with precious substances. Examples like the head-shaped reliquary of St Yrieix or that of Ste Foy (fig. 31) in the form of a seated figure with outstretched arms were encased in sheets of gold inset with precious stones *en cabochon*. In the same way sacred manuscripts were commonly bound in covers enriched with ivory, gold and precious stones as a way of marking and enhancing the reverence in which they were held (fig. 32). Again,

Figure 31 Carolingian reliquary statue of Sainte Foy, Conques. The figure is shaped from sheet gold and studded with amethysts, opals, sapphires, garnets, agates and carnelians set *en cabochon*. Height 850 mm. (Treasury, Ste Foy, Conques)

Figure 32 Cover of the Book of the Pericopes. The cover is of ivory mounted in a gold-covered frame encrusted with enamel plaques and jewels. The elaborately carved ivory was worked in France during the ninth century. The twelve enamel plaques were made in Byzantium during the tenth century, but the roundels were made locally at Regensburg. The book once belonged to Henry II, Duke and King of Bavaria and later (1014–24) Holy Roman Emperor, but was presented to the cathedral at Bamberg at its dedication in 1012. (Munich, Staatsbibliothek)

90

altar fronts like those of St Ambrose, Milan, and St Mark, Venice, along with others represented in early paintings, were covered in embossed sheet gold and enriched with precious stones and enamels. The ritual vessels placed on altars conformed to the same tradition. Abbot Suger's famous two-handled cup (fig. 33) dating from c. 1140 incorporated an antique body carved from sardonyx. The frame is studded with pearls and precious stones. Even humble parish churches possess in many cases silver treasures in the paten and chalice used in the rite of communion. Again, the symbolic value of the cross was heightened by the application of gold or silver. In grand surroundings like the choir of St Denis the golden cross made around 1140 stood twenty-three feet high. The goldsmiths employed by abbots and bishops as well as by kings embellished gospel covers by enclosing them in gold set with gems, just as scribes illuminated sacred texts in gold leaf. The high role of priests was symbolized by the precious substances lavished on their vestments and appurtenances. The heads of a bishop's crozier and staff were often carved from ivory (Plate G), as were the liturgical combs used by officiants at the altar on high occasions, and the episcopal ring would be made of gold and set with a precious stone.

The contribution to religious rites made by precious substances, though based on their appeal to the senses and enhanced by appreciation of their value, was deepened and enriched by symbolism. The breastplate worn by High Priests of Israel to contain the sacred lot is described in Exodus[24] as being set with four rows of precious stones, each engraved with the name of one of the tribes of Israel. Among these were many of those most favoured in the ancient world, including agate, carnelian, jasper, lapis lazuli, sard and turquoise. The Medieval church went to some lengths to specify the roles of particular stones in religious imagery. Gregory the Great found time to designate the varieties most appropriate to the different orders of the Hierarchy of Heaven – sard for Seraphim, topaz for Cherubim, jasper for Thrones, chrysolite for Dominions, onyx for Principalities, beryl for Powers, sapphire for Virtues, carbuncle for Archangels and emerald for Angels.[25] Again, pearls were extensively used in early Christian iconography as symbols of regeneration. Enough has been said even in this brief review to make the point that when men wished to designate sanctity or degrees of sanctity it was to the hierarchy of precious substances that they turned for appropriate symbols.

High ecclesiastics were men of authority as well as of sanctity. Apart from presiding over their own hierarchies they might also, as they most certainly did in Medieval Christendom, play important roles in the wider community of the state. Like kings, bishops sat on thrones. At times when the generality stood in the presence of their superiors and at other times squatted on the ground or at

Figure 33 Chalice of Abbot Suger, c. 1140, from St Denis, France. The cup itself is formed from an antique sardonyx vessel. It is now held in a gilt frame mounted with pearls and precious stones *en cabochon*. (National Gallery of Art, Washington, D.C.)

most sat on stools or benches, thrones were potent symbols of authority. As such they might be graded in height to match variations in status, just as they still are in respect of sovereign and consort at state openings of the British Parliament. As symbols wooden thrones have traditionally attracted some of the most precious substances. The back of Tutankhamun's throne was covered in sheet gold, and according to the Bible King Solomon's (970–933 B.C.), mounted on a dais with six steps, was sheathed in ivory overlaid by gold. A fragment of one belonging to Sargon II of Assyria (722–705 B.C.) in the British Museum is encrusted in gold, ivory, carnelian and lapis lazuli. In the Mediterranean ivory was the most favoured precious substance used to dignify and embellish wooden thrones. The consuls and high magistrates of Republican Rome sat in folding curule chairs inlaid with ivory. The earliest chair of St Peter at Rome was enriched by ivory, and Archbishop Maximian's throne at Ravenna, made in 546, was sheathed in ivory plaques (fig. 34). The Peacock Throne taken by Nadir Shah at the sack of Delhi in 1739 and later removed to Persia embodied the strongly developed Mogul taste for precious stones. In recent times gilt has been widely used to create the effect of gold, but during the seventeenth and eighteenth centuries there was a fashion in Europe for silver thrones, like that made for the coronation of the Queen of Denmark in 1731.

Precious substances contributed most effectively to the functioning of traditional societies by defining roles in the functioning of the hierarchy of authority. This was particularly important when few persons were able to read and had to depend on what they were told and above all on what they could see. The utmost importance was attached to the rituals and paraphernalia of government and not least to the artefacts that made visually explicit the status of the leading actors. The focal symbol of sovereignty is the crown set on a sovereign's head, the seat of perception and decision, during the rite of coronation. The notion of designating sovereigns by encircling their heads with golden crowns set with precious stones was already ancient when Pepin was crowned King of the Franks at Soissons in 751. In western as in eastern Christendom coronation was a religious rite which began with the anointment of the ruler with holy oil. Those present at the coronation of Pepin must have known, if only from the Old Testament, that the rite in which they took part was already one of great antiquity. When David overcame the Ammonites, we are told 'he took the crown of the king off his head and found it to weigh a talent of gold and there were priceless stones in it; and it was set on David's head'.[26] Coronation was a religious rite, but it also served the political purpose of legitimizing the sovereign (figs. 35–7).

By making crowns of the most precious metal, gold (and in the case of that made for Queen Alexandra platinum), and setting them with the most resplen-

Figure 34 Archbishop Maximian of Ravenna's ivory-covered throne, A.D. 546–7. The throne is thought to have been made in Constantinople and given by Justinian to his viceroy Maximian. Carved on the front is the bishop's monogram. (Museo Arcivescovile, Ravenna)

94

Figure 35 St Stephen's Crown, the Holy Crown of Hungary. The circlet is enriched by enamel plaques of Byzantine manufacture, alternating on the lower register with jewels *en cabochon*. The arches have been inset with pearls and uncut diamonds. Dated 1074–7. (Budapest Treasury)

95

dent stones available, a combination of the most precious substances proclaimed the supremacy of the state and its titular head.[27] Crowns were subject to frequent reconstruction. British monarchs up to and including George IV were often too hard up to maintain their crowns fully jewelled. The usual thing was to hire diamonds from the court jeweller for the coronation and then return them.[28] What passed on were the gold frame and certain historic jewels from earlier dynasties. If sovereignty was most obviously epitomized in the most splendid regalia, the incorporation of ancient components was a useful way of strengthening legitimacy. When Richard Earl of Cornwall was crowned Holy Roman Emperor in 1257 he found it useful to have cameos and intaglios dating from Classical antiquity inserted in his crown as well as precious stones (fig. 36). Even Napoleon found it expedient to buttress force by prescription, having antique gems implanted in his crown and terming it the crown of Charlemagne.

When Ivan IV had himself crowned as Tsar or Caesar of All the Russias in 1547, he took care to combine the claims of his predecessors as Grand Dukes of Muscovy to be Protectors of Orthodoxy, following the fall of Constantinople, with his own position as heir to the territorial gains made by his immediate pre-

Figure 36 The crown of gilded silver worn by Richard, Earl of Cornwall, for his coronation as Holy Roman Emperor at Aachen in 1257. A crocketed arch running from front to back originally supported a mitre on the inside. Diameter 210 mm. (Cathedral Treasury, Aachen)

decessors Ivan III and Vasily III. Ivan's coronation is particularly relevant to our theme because we have an eye-witness account of the spectacle he and his Empress presented immediately afterwards.[29] The Tsar, we are told,

was taken out of his chair of majestie, having upon him an upper robe adorned with precious stones of all sorts, orient pearls of great quantitie, but always augmented in riches. It was in weight two hundred pounds; the traine and parts thereof borne up by six dukes. His chief imperiall crowne upon his head, very precious; his staff imperiall in his right hand, or an unicornes horne of three foot and a halfe in length, beset with rich stones.

Before him were carried his 'sceptre globe . . . his rich cap, beset with rich stones and pearles . . . and his six crownes'. As he came to the door he found his horse ready 'most richly adorned with a covering of imbrodered pearle and precious stones, saddle and all furniture agreeable to it'. In the meantime his lady was exposed to public gaze seated in her chair at a great open window. 'Most precious and rich were her robes, and shining to behold, with rich stones and orient pearle beset.' The effect of the display of these quantities of pearls and precious stones was clearly designed to intensify public devotion, as Queen Elizabeth I of England was well aware (Plate J).

When Peter the Great took the final step of proclaiming himself Emperor in 1721 he discarded the jewelled cap in which he and his brother Ivan had originally been crowned in 1682. In its place he adopted a mitre crown of the type introduced by the Holy Roman Emperor Frederick III in 1442. Catherine II persisted with this form but had it more richly jewelled (fig. 37). In the event her crown was set with no less than 4,936 diamonds and her sceptre was headed by the Orloff diamond, a stone of 194.5 carats. The wealth of diamonds incorporated in her regalia reflected the fact that Catherine presided over a realm that for a short period was the world's leading producer. In addition the large and perfectly matched pearls outlining the margins of the mitre crown and the balas ruby of 414.3 carats mounted at its apex were enough to proclaim her ready access to the treasures of the orient. Her orb, made in haste for the occasion by a French jeweller, incorporated precious stones from many sources. Brazilian diamonds were used to define the main bands, a fine Indian diamond was set at the middle, a Ceylonese sapphire appeared above the globe and over all was set a cross of pink diamonds. Regalia of such splendour symbolized Catherine's European standing as a despot, a species which however benevolent it might declare itself was destined for a short life. Her crown, in some respects the most resplendent, proved in fact to be the last ever made for a Russian ruler. It was most recently worn at the State Opening of the Duma in 1906. The regalia of Russian tsardom are now displayed in the Kremlin as relics of a bygone age.

By the poetry of history the crown of the British sovereign, who ceased to rule in all but name in the aftermath of the Glorious Revolution of 1688, is the only one of consequence, the papal tiara aside, still used in the rite of coronation. Because British crowns were and happily still are used they have continued to respond to changed circumstances. The royal regalia have above all reflected and symbolized the extension of the Empire to embrace territories rich in diamonds. Queen Victoria celebrated the transference of power in India from the Company to the Crown in 1858 by accepting the Koh-i-Noor, one of the most splendid diamonds accumulated by the Mogul emperors, a stone at present mounted at the centre of the cross in front of Queen Elizabeth the Queen mother's crown. When the Cullinan diamond of 3,106 carats was recovered from the Premier Mine, Pretoria, its size and quality ensured at that time only one destination, the regalia of the king-emperor. The largest of the nine stones into which it was shaped, the largest Star of Africa (530.2 carats), was mounted at the head of the Sceptre with the Cross, the second, a cushion-shaped stone of 317.4 carats, was

Figure 37   Catherine II's mitre crown, the Imperial Crown of Russia. (Kremlin Armoury, Moscow)

98

incorporated in the front of the Imperial State Crown (see Frontispiece), displacing the Stuart sapphire, and the third and fourth were mounted in the consort's crown. Along with immensely valuable diamonds the regalia incorporate stones from earlier stages in the history of the monarchy. The Imperial State Crown includes the huge balas ruby, acquired by the Black Prince and worn by Henry V at Agincourt, mounted at the heart of the cross above the second Star of Africa. The Stuart sapphire, once in front, is now mounted in the rear of the crown. This stone itself had an intriguing history. Taken to France by James II on his flight, it was later worn by the Cardinal Duke of York on his mitre before coming home again to Queen Victoria by way of George IV. The State Crown of the House of Windsor thus incorporates precious stones from the Plantagenets and Stuarts, alongside the spoils of India and the Rand. In this way it combines the legitimacy conferred by long inheritance with the lustre and potency of some of the most prestigious diamonds recovered by man.

The fate of the Indian diamonds which found their way to the court of France was less assured. The point may be illustrated by tracing the fate of two of the

Figure 38 The Sceptre with the Cross, a component of the British Crown Jewels, incorporating the Star of Africa, at 530.20 carats the largest cut diamond in the world (Crown Copyright).

stones stolen at the time of the French Revolution. The Regent diamond of 410 carats recovered from Golconda in 1701 was soon acquired by Governor Pitt, grandfather and great-grandfather of men who between them despoiled a great part of the French empire. Pitt sold his prize in 1717 to the Duke of Orleans, Regent of France during the boyhood of Louis XIV. Stolen early on in the Revolution, the Regent diamond was recovered by the French Adjutant-General and pawned first to a German banker and then to a Dutchman to secure loans. The stone next appeared in the hilt of the sword worn by Napoleon at his coronation. When the great man was exiled to Elba the diamond was returned to his father-in-law, Francis I of Austria, who duly handed it back to the French authorities. It now reposes in the Louvre, having escaped the Nazis hidden behind a marble fireplace in the Chateau of Chambord throughout the war. The Great Sancy takes it name from a French ambassador to Turkey who acquired it in Constantinople and on his return to France in 1593 lent it to Henry IV as a pledge for money to pay his troops. While serving as ambassador at the court of St James Sancy's brother sold the stone to James I. When James II fled to France he took refuge at St Germain and sold the stone to Louis XIV. After the Revolution the Sancy diamond was sold by a French merchant to the Russian Demidoff family. We next hear of it being bought in 1865 by Sir Jamsetjee Jeejeebhoy who resold it to a Paris jeweller. The later fate of the Sancy diamond has not been established with complete certainty, but it was probably the stone bought by William Waldorf Astor, 1st Viscount Astor, in 1906 as a wedding present for his son. His daughter-in-law Nancy Astor later wore it in her tiara, a piquant commentary on those who imagined that by overthrowing the ancient regime they were promoting equality.

The production of the symbols of sanctity, majesty and varying degrees of status needed for the effective running of societies based on hierarchy called for highly skilled designers and artificers who must as specialists have been immune from the common round of everyday labour. Celtic chiefs, Anglo-Saxon kinglets and the ecclesiastical dignitaries of medieval and early modern Europe must have maintained their own schools of specialized craftsmen absolved from the common round of labour. Many of these men would have been individuals of vision and creativity as well as manual dexterity. In the early middle ages we find a saint, St Eligius (588–663), acting as goldsmith to Frankish kings. Again, we know that some of the leading artists of Renaissance Italy graduated as apprentices of prominent goldsmiths.[30] The sculptor Benvenuto Cellini, creator of the large Perseus statue at Florence as well as of the famous salt made for Francis I of France, learned his trade from the chief jeweller to the Medici. Dürer, Holbein and Nicholas Hilliard (see Plate J) are among the famous artists of the fifteenth and sixteenth centuries in Germany and England known to have engaged in goldsmith's work. As late as the eighteenth century some of the most renowned painters of France, Boucher, Lancret and Watteau among them, found it no more beneath them to paint gallant scenes on the artificial eggs presented by the king to members of his court at Easter than Cellini did to be called on to unpick the precious stones from Clement VII's tiara as the Imperial troops advanced to sack Rome in 1527. If we may find this incongruous, that is only because we live in an age of collapsed hierarchies in which painters and sculptors are liable to be more highly regarded than their patrons or even their sitters.

Although the substances treated in this work were not categorized as precious primarily on account of their economic value, the mere fact of their being accepted as such, combined with their scarcity in nature, inevitably led to their being used to serve economic ends. Before modern institutions for extending credit were available, insignia and jewels were well adapted to serve as sureties for loans. When Henry V of England was short of cash for the French wars he pawned one of his crowns, probably the Plantagenet crown, to the Mayor of Norwich. Popes were not always above pawning their tiaras. One of the richest made for Julius II later escaped the sack of Rome because it happened to be in pawn at the time. Precious things were also subject to extortion. When Napoleon offered to sign the treaty of Tolentino in 1797, but demanded several million lire to be paid in diamonds and treasure in return, the reigning pontiff was driven to having his tiaras unpicked and handing over among other stones the fabulous emerald collected by Julius II. Only seven years later the pope recovered this stone in the new tiara presented to him to make him look the part while anointing Napoleon as self-crowned Emperor.

Of greater moment, it was to precious substances that men turned for currency, since by definition these possessed the prime quality of acceptability. By no means all such substances qualified in other respects. The most precious lacked the no less vital requirement of uniformity. Diamonds and other highly precious stones varied too greatly in individual quality and worth for use as currency. Cowrie shells suffered from the opposite. They were uniform but occurred in such abundance and could so readily be transported that even when they were taken over as currency they could only serve as a rule for small change. By contrast, metals enjoyed many advantages. They could easily be moulded into standard forms and stamped with denominations and the names of the authorities responsible for issuing them. Also they had the special quality that their purity could be readily controlled. Above all they presented a long established ranking order: gold and silver for higher denominations in descending order and non-precious copper for small change.

The first communities to engage in striking inscribed coins[31] were the trading cities of Lydia, Ionia and Greece during the latter half of the eighth century B.C. (fig. 39: a, b). The first such coins were made from electrum from the rivers that flowed down from the Troilus Mountains, but advantage was soon taken of the different status in the traditional value system of the gold and silver components to separate them and use them for different denominations. From that time the ability to strike coins from the most precious metals became a matter for emulation. By comparison with the Greeks and Etruscans the Romans were slow to accept coinage and when they did so they began with copper. The silver *denarius* was not established on a permanent footing until the Second Pubic War (fig. 39: c, d), and gold only came to dominate the imperial coinage and advertise the power and majesty of Rome with the extension of the Empire (fig. 39: e, f). By the same token the collapse of the Roman Empire led to the cessation of gold coinage in the west, though this was tenaciously delayed in the east. Not surprisingly the restoration of gold coinage became a prime aim of rulers in the states which emerged during the early middle ages on the ruins of the Roman Empire. For centuries the emergent states had to limit their ambitions to silver pennies, successors of the *denarius*. The first English gold coin was struck only in 1255 and it was not until 1344 that the country was able to begin to sustain what proved to be no more than half a millennium of minting a gold denomination in its regular coinage. When the United States regulated its coinage under the legislation of 1792 it displayed independence by designating the silver dollar as the basic unit. Even so the age-old hierarchy prevailed. Gold eagles were struck to save handling so many dollar pieces, and bronze was needed to provide small change. If the monetary authorities of modern consumer societies have had to replace gold by paper in the course of inflation, they are still constrained at moments of crisis to shift ingots of gold from one bank to another, and the price of gold bullion on the exchange remains a sensitive index of confidence in the international market.

The fact that precious metals have so often been minted to serve as coins may serve to emphasize that precious substances are by no means confined to the summits of ecclesiastical or temporal power, even if their most prestigious manifestations were formerly concentrated on these. On the contrary they penetrate many levels of society and enter into a wide range of transactions. The leaders of all human societies, however absolute their authority, are in practice compelled to govern and defend their realms by delegating powers to a wide range of officials, local as well as central, and civil as well as military. Every stage in the growing complexity of social life and in the enfranchisement of populations has emphasized the need to designate functions and grades by means of symbols. The

Figure 39 a, b: The earliest inscribed Greek coin issue known, struck in electrum at Ephesus, Ioni;
around 600 B.C. (× 4).

c, d: Roman *denarius* of c. 209 B.C. showing Roma and the Dioscuri on horsebacl
(× 2.5).

e, f: Roman gold *aureus* of Caesar and Plancus, 45 B.C. (× 2.5). (British Museum) sheer volume of insignia required for public services means that insignia can be given only the appearance of precious metals. Even peers have to be content with coronets of silver gilt, and holders of the orders of chivalry, today awarded for services to the professions, the arts, business and a wide range of public entertainment as well as to politicians and servants of the state, are designated by gilt insignia embellished with enamel. For marking the grades of service officers, braid and metal emblems have to suffice, in Britain gold for the armed services, silver for civilian.

One of the most notable and significant aspects of the more complex societies which emerged in the course of modern history has been the extent to which social life has been moulded by associations of private persons concerned with promoting a wide range of activities from local government, education, the professions, the arts and sports to a widening area of hobbies and pursuits. A common requirement of such bodies is the need for symbols, symbols to embody their own sense of continuity and symbols to encourage and reward the attainment of higher standards. No less universal is their reliance on precious metals to fabricate them. Gold and silver, once reserved for the service of cults and sovereigns, were displayed as plate by corporations like the City of London, the Inns of Court, the City Companies (fig. 40) or the colleges of Oxford or Cambridge and

Figure 40 The buffet plate of the London Goldsmiths' Company.

embodied as medals awarded by academies and other institutions for distinguished contributions to many fields of learning and professional activities (Plate K). The enfranchisement of large populations and the consequent rise of mass spectator sports and popular entertainment have served only to widen and intensify the desire to excel. The gold and silver cups put up by aristocrats and wealthy persons as prizes for horse races (fig. 41) are now matched by trophies competed for by teams in a bewildering variety of sports (fig. 42). Such trophies attract the kind of devotion from supporters once accorded to regalia and the furnishing of shrines. The medals awarded at each Olympic Games are of uniform design. Merit is defined by the hierarchy of metals from which they are struck: gold for the winner, silver for the second best and bronze for the third (fig. 43).

The social transformation brought about by the enfranchisement, economic as well as political, of working populations has not impaired the regard for precious substances as symbols of excellence in every field of endeavour, any more than it lessened the hunger for personal jewellery remarked in our opening chapter. The more democratic communities become, it would seem, the keener the anxiety of

Figure 41 The Derby Trophy

Figure 42 The Football Association Cup

individuals to excel their neighbours. The conclusion seems evident. Regardless of social circumstances a concern for excellence and a desire to recognize it through symbols embodying precious substances is widespread among our species. To judge from the surviving traces, this applies more especially to communities which in the course of the last five millennia have dragged themselves from the morass of primitive communism and set their feet on ground firm enough to support civilized ways of life. However diverse the civilizations embodied in the archaeological record, they reveal clear signs of having recognized hierarchies of excellence. If one of the necessary chores of an archaeologist is to reconstruct subsistence and technology, it is one of his chief privileges to explore the values of the civilizations coming within his purview and not least the symbols which have served to define status and acknowledge outstanding achievement.

Figure 43 Olympic silver medal designed in the Art Nouveau style for the London 1908 Olympiad by the Australian Bertram Mackennal (1863–1931). Struck silver, 33.5 mm in diameter.

Notes on the colour plates

Frontispiece The Imperial State Crown. When this crown was designed for Queen Victoria's coronation in 1837, the Stuart sapphire was placed in the front of the circlet. When King Edward VII incorporated the Second Star of Africa, cut from the Cullinan diamond, in this position, the Stuart sapphire was re-mounted in the rear of the circlet. The circlet also carries notable emeralds and sapphires as well as many small diamonds. The Black Prince's balas ruby is mounted in the middle of the cross above the Second Star of Africa and some fine pearls are visible on the edges of the arches of the crown. The symbolic use of diamond-studded fleurs-de-lys was retained by George IV for his coronation crown in 1820, despite the fact that the British claim to the crown of France had been given up at the time of the Treaty of Amiens nineteen years previously. (Crown Copyright)

Plate A

Amber objects from Mesolithic Denmark. Note the range of colours. The miniature brown bear is from Resen on Jutland and the silhouette elk head from Egemarke, West Zealand. The accompanying pieces are simple pendants, drilled at the top for suspension; the ornament consists of surface pricking and simple linear motifs engraved in rows or panels. The length of the bear is 70 mm. (Nationalmuseet, Copenhagen)

Plate B

Maori adze with nephrite blade, designed for use as a symbol of chiefly authority. The elaborate carving of the haft proclaims its ceremonial status. (Museum of Mankind, London)

Plate C

Egyptian Middle Kingdom necklace, said to be from Thebes, about 1900–1800 B.C. The six large beads and clasp imitating cowrie shells, as well as the pendant beards and fish and some of the smaller spherical beads, are made of electrum. Others of the small beads are made from amethyst, carnelian, green felspar and lapis lazuli. The small lotus flower pendant is inlaid with carnelian and blue and green glass. Overall length 463 mm. (British Museum)

Notes on the colour plates

Plate D

Sumerian court jewellery from the royal cemetery of the Third Dynasty at Ur in modern Iraq, about 2500 B.C. Note particularly the thinly hammered gold of the hair ornaments and the abundant use of lapis lazuli and carnelian. Height of model 650 mm. (British Museum)

Plate E

Mosaic mask from Mexico. This has commonly been accepted as representing the Toltec god-king Quetzalcoatl but it might, alternatively, represent the sun-god Tonatiuh. The turquoise mosaic, the pearl shell eyes and the white shell teeth have all been applied to the wooden matrix by a resinous gum. The circular holes through the eyes suggest that the mask was intended to be worn, at least on occasion, by a priest taking part in ceremonies connected with the god. (Museum of Mankind, London)

Plate F

Maori neck ornament (*tiki*) carved from greenstone. Early nineteenth century. Height 175mm. (Cambridge University Museum of Archaeology and Anthropology)

Plate G

Romanesque bishop's crozier carved from walrus ivory and dated stylistically to the late eleventh century by comparison with manuscript illuminations. The incident represented may be one recorded by Bede (*Hist. Ecc.* V, 2), the healing of a dumb youth by St John of Beverley, Bishop of Hexham (686–705). St John carries a cross while an attendant physician in the foreground inspects the youth's afflictions. Height 96 mm. (Private collection, on loan to the British Museum)

Plate H

Tutankhamun's innermost coffin of pure gold (detail). This was enclosed by two gilded wooden coffins and a stone sarcophagus and contained the mummy with its gold portrait-mask overlain by a whole network of jewels and amulets. The coffin itself is of 22 carat gold 2.5–3 mm thick, inlaid with multi-coloured glass paste and semi-precious stones. It depicts the mummified figure of the king as Osiris, his arms crossed on his chest, holding the characteristic insignia of flail and shepherd's crook. Length of sarcophagus 1880 mm. Weight 110.4 kilograms. (Egyptian Museum, Cairo)

Notes on the colour plates

Plate I

Gold pectoral from the tomb of Tutankhamun. Cloisonné work inset with semi-precious stones and glass paste; the central motif consists of a falcon with upward-curving wings representing the god Horus, his body and head formed from a fine chalcedony scarab, symbol of resurrection. The front claws and tips of the wings support a ship carrying two symbols of the moon: the left *udjat* eye and the mooon's disc with crescent. Gold figures of Tutankhamun flanked by Ra-Harakhty (a fusion of the sun god Ra and the sky god Horus) and the moon god Thoth have been applied to the silver disc. Length 510 mm. (Egyptian Museum, Cairo)

Plate J

The 'Phoenix' portrait of Queen Elizabeth I, probably by Nicholas Hilliard around 1575–80. This panel, called after the Phoenix jewel at the breast, emphasises the importance of costume in the display of majesty. The queen is shown wearing a pearl girdle and a jewelled collar and necklace displaying diamonds and rubies, the latter with a huge sapphire in addition. 775×610 mm. (National Portrait Gallery, London)

Plate K

H.R.H. Prince Philip, Chancellor of the University of Cambridge, wearing academic robes enriched with gold lace to signify his place at the head of the academic hierarchy. The robes, of a design that has remained the same since the seventeenth century at least, are of silk satin damask embellished with gold plate lace and gold wire ornaments bearing rosettes, olivettes and fringes. At the sleeve heads are wings worked in a traditional design of oakleaves and acorns. The making time for the ornaments and wings is around 320–50 working hours, whereas the cutting and hand-sewing of the robes themselves will take an experienced worker only an additional 120 working hours. Eight metres of silk damask, thirteen metres of gold plate lace and some seventy gold wire ornaments were used in these particular robes, the work of Ede and Ravenscroft Ltd of Chancery Lane.

Notes

1 Introductory

1 Grahame Clark, *The Identity of Man* (Methuen, London, 1982).

2 Colin Renfrew, *Before Civilization* (Cape, London, 1973), 187.

3 William Watson, *Ancient Chinese Bronzes* (Faber, London, 1962); Wen Fong (ed.), *The Great Bronze Age of China* (Thames and Hudson, London, 1980).

4 Heather Lechtman, 'Andean Value Systems and the Development of Prehistoric Metallurgy', *Technology and Culture*, 25 (1) (1984), 1–36.

5 Thorstein Veblen, *The Theory of the Leisure Class*. Originally published by Macmillan, N.Y., in 1899. References are taken from the edition by Unwin Books, London, 1970, with introduction by Prof. C. Wright Mills.

6 *ibid.*, 85.

7 *ibid.*, 96.

8 *ibid.*, xiv.

9 A. Leroi-Gourhan, *The Dawn of European Art* (Cambridge University Press, 1982), 43.

10 A. Marshack, *The Roots of Civilization* (McGraw-Hill, N.Y., 1972).

11 W. Buttler, 'Beiträge zur Frage des jungsteinzeitlichen Handels', *Marburger Studien* (1938), 26–33.

12 E. S. Higgs and C. Vita-Finzi, 'Prehistoric Economies: A Territorial Approach', *Papers in Economic Prehistory*, ed. E. S. Higgs (Cambridge University Press, 1972), 27–36.

13 W. J. Sollas, *Ancient Hunters* (3rd edn, London, 1929), 588.

14 S. C. Nott, *Chinese Jade Throughout the Ages. A Review of its Characteristics, Decoration, Folklore and Symbolism* (London, 1936), 10.

15 Bo Gyllensvärd, *Chinese Gold and Silver in the Karl Kempe Collection* (Stockholm, 1953); *Chinese Gold, Silver and Porcelain in the Kempe Collection* (The Asia Society, N.Y., 1971).

16 W. F. Foshag, 'Chalchihuitl – A Study in Jade', *American Mineralogist*, 40 (1955), 1066.

17 Joan Evans, *A History of Jewellery, 1100–1870* (London, 1953), 202.

110

2 Organic materials

1 T. K. Penniman, *Pictures of Ivory and Other Animal Teeth, Bone and Antler* (Occ. Papers Techn. Pitt-Rivers Museum Oxford, 1953) 13; see also G. C. Williamson, *The Book of Ivory* (Frederick Muller, London, 1938).

2 T. G. H. James, *An Introduction to Ancient Egypt* (British Museum Publ., London, 1979), 13, fig. 51, and 234.

3 R. D. Barnett, *Ancient Ivories in the Middle East* (Qedena Monographs, no. 14, Hebrew University, Jerusalem, 1982); 'Fine Ivory Work', *A History of Technology*, ed. C. Singer (Oxford, 1954), ch. 24; Williamson, *op. cit.* (1938), 46–60.

4 H. Obermaier, *Fossil Man in Spain* (New Haven, 1925); E. A. Golomshtok, *The Old Stone Age in European Russia* (Am. Phil. Soc., Philadelphia, 1938); P. P. Ephimenko, *Kostienki I* (Moscow, 1958); I. G. Shovkoplyas, *Mezinskaya Stoyanka* (Kiev, 1965); C. B. M. McBurney, *Early Man in the Soviet Union. The Implications of Some Recent Discoveries* (British Academy, 1976).

5 C. B. M. McBurney, *The Stone Age of Northern Africa* (Penguin Books, London, 1960), 258–72 and pl. 17–19; J. D. Clark, *The Prehistory of Africa* (Thames and Hudson, London, 1970), 174.

6 Barnett, *op. cit.* (1954), 664 and fig. 455.

7 Livy, *History of Rome*, XXXVII, 58.

8 D. Talbot Rice, *The Art of Byzantium* (Thames and Hudson, London, 1959), 291 and fig. 19.

9 William Watson (ed.), *Chinese Ivories from the Shang to the Qing* (Oriental Ceramic Society, London, 1984).

10 R. E. M. Wheeler, *Rome Beyond the Imperial Frontiers* (Penguin, London, 1955), 192ff.

11 J. J. L. Duyvendak, *China's Discovery of Africa* (Probsthain, London, 1949); G. S. P. Freeman Grenville, *The East African Court: Select Documents from the First to the Earlier Nineteenth Century* (Oxford, 1962); Hikoichi Yajima, 'Maritime Activities of the Arab Gulf People and the Indian Ocean World in the Eleventh and Twelfth Centuries', *J. Asian & African Studies*, 14 (1977), 195–208.

12 H. Carter and A. C. Mace, *The Tomb of Tut-ankh-amen*, vols. 1–3 (Cassell, London, 1923–33); Williamson, *op. cit.* (1938), 120ff.

13 H. L. Lorimer, *Homer and the Monuments* (Macmillan, London, 1950), 61–3; J. B. Wace and Frank H. Stubbings (ed.), *A Companion to Homer* (Macmillan, London, 1962), 533; G. Loud, *The Megiddo Ivories* (Orient. Inst. Publ. 52, Chicago, 1952).

14 Barnett, *op. cit.* (1954), fig. 459.

15 L. D. Caskey, 'A Chryselephantine Statuette of the Cretan Snake Goddess', *American J. Archaeology*, 1915, 237–49.

16 Barnett, *op. cit.* (1954), 661.

17 O. M. Dalton, *Catalogue of the Ivory Carvings of the Christian Era* (British Museum, London, 1909).

18 Bo Gyllensvärd, *Chinese Art from the Collection of H.M. King Gustaf VI Adoloph of Sweden* (Exhib. Cat. British Museum, 1972), 22–4 & nos. 182–99; 'King Gustaf VI Adolph's Approach to Chinese Art', *Trans. O.C.S.* (1978–80), 31–34, esp. 42–3.

19 *The Times*, 29 March 1984.

20 *Chinese Treasures from the Avery Brundage Collection* (The Asia Soc. Inc., New York, 1968), pl. 9; Wen Fong (ed.), *The Great Bronze Age of China*, pl. 93.

21 R. Soame Jenyns, *Chinese Art. The Minor Arts*, II (Oldbourne Press, London, 1965), ch. 3.

22 Duyvendak, *op. cit.*

23 W. J. Sollas, *Ancient Hunters* (3rd edn, London, 1929), 588.

24 Hugh Tait (ed.), *Jewellery Through 700 Years* (British Museum, 1976), no. 339 & colour pl. 6.

25 C. H. V. Sutherland, *Gold. Its Beauty, Power and Allure* (Thames and Hudson, London, 1959), pl. 4.

26 K. Kenyon, 'Early Jericho', *Antiquity*, 35, 1–9.

27 H. Kelm, *Kunst vom Sepik*, Band 1 (Museum für Volkerkunde, Berlin), no. 281.

28 J. Wilfrid Jackson, *Shells as Evidence of the Migrations of Early Culture* (Manchester, 1917), 158, 174, 176.

29 Jessica Rawson, *Ancient China* (British Museum Publ., 1980), 101.

30 Wen Fong (ed.), *The Great Bronze Age of China*, 42.

31 Rawson, *op. cit.*, 180.

32 *ibid.*, 99.

33 Kwang-chih Chang, *The Archaeology of Ancient China* (rev. edn, Yale, 1968), 432.

34 Jackson, *op. cit.*, 183.

35 Alan F. C. Ryder, *Benin and the Europeans 1485–1897* (Longman, London, 1969), 61.

36 B. Fagg, 'Archaeological Notes from Northern Nigeria', *Man*, 46 (1946), 52.

37 Jackson, *op. cit.*, 131.

38 Hjalmar Stolpe, 'Sur les découvertes faites dans l'ile de Björkö', *Congr. int. d'Anthr. et d'Archéol. Préhist. C.R. 7me sess. Stockholm 1874*, 613–31, esp. p. 626.

39 *Victoria County History of Cambridgeshire*, vol. 1, 294, fig. 26.

40 Ryder, *op. cit.*, 299.

41 *ibid.*, 292.

42 Ling Roth, *Great Benin. Its Customs, Art and Horrors* (F. King and Sons, Halifax, 1903), 26f.

43 G. C. Williamson, *The Book of Amber* (Benn, London, 1932); John Munro, *Aramco World Magazine*, Nov.–Dec. 1981, 33–6.

44 Bo Gyllensvärd, *op. cit.* (1953), 155 and fig. 15.3.

45 A. Rust, *Das Altsteinzeitliche Rentierjäger Meiendorf* (Neumünster, 1937), 111ff, taf. 55.

46 J. G. D. Clark, *Excavations at Star Carr* (Cambridge, 1954), 165.

47 Therkel Mathiassen, *Danske Oldsager. I. Aeldre Stenalder* (Copenhagen, 1948),

nos. 227–9 and 231–3; Grahame Clark, *The Earlier Stone Age Settlement of Scandinavia* (Cambridge, 1975), 147, 155, 157f.

48 M. Gimbutas, *The Prehistory of Eastern Europe*. Pt 1 (Am. School Prehist. Res. Bull. 20. Harvard, 1956), fig. 108, pls. 30–3.

49 A. W. Brøgger, *Den Artiske Stenalder i Norge* (Kristiania, 1909), figs. 110–11.

50 J. L. Giddings, *Ancient Men of the Arctic* (Knopf, N.Y., 1967), 320.

51 C. J. Becker, *Mosefunde Lerkar fra Yngre Stenalder* (Copenhagen, 1948), 295–302; J. Brøndsted, *Danmarks Oldtid* (Copenhagen, 1957), I, 186f, 189, 212, 258, 333.

52 T. Wiślański (ed.), *The Neolithic in Poland* (Warsaw, 1970), 441.

53 J. M. de Navarro, 'Prehistoric Routes between Northern Europe and Italy defined by the Amber Trade', *Geog. J.* 66 (1925), 481–507.

54 J. G. D. Clark, *Prehistoric Europe: The Economic Basis* (Methuen, 1952), 261–6.

55 G. von Merhart, 'Die Bernsteinschieber von Kakovatos', *Germania*, 24 (1940), 99–102.

56 B. Laufer, 'Historical Jottings on Amber in Asia', *Mem. Am. Anthrop. Assoc.*, vol. 1, pt 2 (1907), 215–44.

57 *Chambers's Encyclopaedia* , 3rd edn (1950), vol. 8, 87a.

58 F. Elgee, *Early Man in North-East Yorkshire* (Bellows, Gloucester, 1930), 108–18.

59 Henrietta Davies, *Arch. J.*, 93 (1936), 200ff.

60 S. Piggott, *Neolithic Cultures of the British Isles* (Cambridge, 1954), 176, 311, 361 and fig. 63, no. 2.

61 V. G. Childe, *The Prehistory of Scotland* (London, 1935), 104f.

62 S. Piggott, 'The Early Bronze Age in Wessex', *Proc. Prehist. Soc.* 4 (1938), 79.

63 Joan Taylor, *Bronze Age Gold Work of the British Isles* (Cambridge, 1980), 25.

64 Elgee, *op. cit.*, 117.

3 Jade

1 For a brief survey, see Elisabeth H. West, 'Jade: its character and occurrence', *Expedition* (1963), 2–11. Philadelphia.

2 W. F. Foshag, 'Chalchihuitl – a Study in Jade', *American Mineralogist*, 40 (1955), 1062–70.

3 For tables, see S. C. Nott, *Chinese Jade Throughout the Ages* (London, 1936).

4 Raymond Firth, *Primitive Economics of the New Zealand Maori* (London, 1939), ch. 12, esp. p. 388.

5 L. H. Fischer, *Nephrit und Jadeit*. 2nd edn, Stuttgart, 1880.

6 A. B. Meyer, 'Jadeit – und Nephrit – Objekte', *Publ. des Ethnographischen Museums Dresden*. Leipzig, 1883–3; 'The Nephrite Question', *American Anthropologist*, 1 (1888), 231–42.

7 W. Campbell Smith, 'Jade Axes from Sites in the British Isles', *Proc. Prehist. Soc.* (1963), 133–72; 'The Distribution of Jade Axes in Europe', *ibid.* 31 (1965), 25–33.

8 Foshag, *op. cit.*, 1069.

9 Grahame Clark, 'Traffic in Stone Axe and Adze Blades', *Economic History Review*, 18 (1965), 1–29, esp. p. 22.

10 West, *op. cit.*, 5 & 7; James C. Y. Watt, *Chinese Jades from Han to Ch'ing* (The Asia Society, N.Y., 1980), 26–30.

11 T. W. Beale, 'Early Trade in Highland Iran: A View from a Source Area', *World Archaeology*, 5 (1973), 133–48.

12 Kwang-chih Chang, *The Archaeology of Ancient China* (rev. edn, Yale, 1971), 199.

13 Qian Hao, Chen Heyi and Ru Suichu, *Out of China's Earth* (Frederick Muller Ltd, London, 1981), 9–27.

14 Jessica Rawson, *Ancient China* (British Museum Publications, 1980), 32.

15 W. Aurel Stein, *Sand-Buried Ruins of Khotan* (London, 1903), 207, 252ff.

16 M. Gimbutas, 'Borodino, Seima and their Contemporaries', *Proc. Prehist. Soc.* 22 (1956), 15f.

17 S. Howard Hansford, *Chinese Jade Carving* (London, 1950), 43–7.

18 West, *op. cit.*, 8; Chester S. Chard, *Northeast Asia in Prehistory* (Univ. Wisconsin Press, 1974), 189, figs. 5, 18; Roderick Whitfield (ed.), *Treasures from Korea* (British Museum Publ., London, 1984), 77 and cat. no. 52.

19 Clark, *op. cit.* (1965), 23.

20 Quoted from Hansford, *op. cit.*, 38.

21 The *T'ien Kung K'ai wu* by Sung Ying-hsing, first published in 1637. Quote from Hansford, *op. cit.*, 39f.

22 Stein, *op. cit.*, 252ff.

23 Watt, *op. cit.*, 27.

24 Hansford, *op. cit.*

25 C. T. Loo, *Chinese Archaic Jades* (N.Y., 1950); S. Howard Hansford, *Jade. Essence of Hills and Streams* (London, 1969), 31–90.

26 John Ayers and Jessica Rawson, *Chinese Jade Throughout the Ages* (Oriental Ceramic Society, London, 1975), nos. 423–4.

27 See the many illustrations in Loo, *op. cit.*

28 Rawson, *op. cit.*, 36ff.

29 *ibid.*, 83.

30 Nott, *op. cit.*, ch. 4, 27–30; Elizabeth Lyons, 'Chinese Jades', *Expedition*, 20 (1978), 12, fig. 13.

31 Qian Hao *et al.*, *op. cit.*

32 Hansford, *op. cit.*, 1950, 121; and *Chinese Carved Jades* (Faber and Faber, London, 1968), 44; Nott, *op. cit.*, 113ff, fig. 60.

33 A. Salmony, *Chinese Jade Through the Wei Dynasty* (N.Y., 1963), 113ff.

34 J. J. M. de Groot, *The Religious System of China*, vol. I, pt 1 (reprinted Taipei, 1964), 272.

35 *ibid.*, 279.

36 B. Laufer, *Jade. A Study in Chinese Archaeology and Religion* (Chicago, 1912, and N.Y., 1974), pl. xxxvi; Rawson, *op. cit.*, fig. 177.

37 Laufer, *op. cit.*, pl. XXXVII, 6–8 and pl. XXXVIII, 1–3.

38 Qian Hao *et al.*, *op. cit.*, 130.

39 Ayers and Rawson, *op. cit.*; Watt, *op. cit.*; and Brian Morgan, *Dr. Newton's Zoo. A Study of Post-Archaic Small Jade Carvings* (Bluett, London, 1981).

40 Campbell Smith, *op. cit.*, fig. 1.

41 T. G. E. Powell, *Prehistoric Art* (Thames and Hudson, London, 1966), fig. 114.

42 John Coles *et al.*, 'A Jade Axe from the Somerset Levels', *Antiquity*, 48 (1974), 216–20.

43 Powell, *op. cit.*, fig. 115.

44 Clark, *op. cit.* (1965); Firth, *op. cit.*; W. T. L. Travers, *Trans. N.Z. Inst.*, 5, 19ff.

45 West, *op. cit.*, 8f.

46 Giddings, *op. cit.*, 306.

47 *ibid.*, 46f.

48 Philip Drucker, Robert F. Heizer and Robert J. Squier, *Excavations at La Venta, Tabasco, 1955* (Bur. Am. Ethno., Bull. 170. Washington, 1959), 133–94.

49 *Handbook of the Robert Woods Bliss Collection of Pre-Columbian Art* (Dumbarton Oaks, 1963), cat. nos. 162–86.

50 Norman Hammond, 'Preclassic to Postclassic in northern Belize', *Antiquity*, 48 (1974), 185.

51 T. Proskouriakoff, 'Jades from the Cenote of Sacrifice, Chichen Itza, Yucatan', *Peabody Mus. Mem.* 10, no. 1, 1974.

52 Richard E. W. Adams, *Prehistoric Mesoamerica* (Boston, 1977), 154.

53 Peter Harrison, 'A Jade Pendant from Tikal', *Expedition*, 5 (1963), 12–13.

54 Quoted by Adams, *op. cit.*, 80, from the English translation by Charles E. Dibble and Arthur J. C. Anderson of Book II, 222, of Father de Sahagun's *General History*.

4 Precious metals

1 C. H. V. Sutherland, *Gold. Its Beauty, Power and Allure* (Thames and Hudson, London, 1959).

2 *ibid.*, 19.

3 Warwick Bray, *The Gold of El Dorado* (London, 1978), 27f.

4 H. Maryon and H. J. Plenderleith, 'Fine Metal-Work', *A History of Technology*, ed. C. Singer (Oxford, 1954), 650.

5 Sutherland, *op. cit.*, 11.

6 D. Jacobsson, *Fifty Golden Years of the Rand, 1886–1936* (London, 1936).

7 A. Tobler, *Excavations at Tepe Gawra*, II (Philadelphia, 1950).

8 Maryon and Plenderleith, *op. cit.*, 624.

9 A. Lucas, *Ancient Egyptian Materials and Industries* (rev. J. R. Harris) (Edward Arnold, London, 1962), 224–35; Sutherland, *op. cit.*, ch. 2.

10 D. E. Strong, *Greek and Roman Gold and Silver Plate* (Methuen, London, 1966).

11 A. Hartmann, 'Prähistorische Goldfunde aus Europa', *Studien zu den Anfangen der*

Metallurgie, vol. 3; Joan Taylor, *Bronze Age Goldwork of the British Isles* (Cambridge, 1980).

12 John Brailsford, *Early Celtic Masterpieces from Britain in the British Museum* (British Museum, London, 1975), chs. 5 and 6.

13 M. B. Mackeprang, *Die nordische Goldbrakteater* (Aarhus, 1952).

14 W. Holmqvist, *Germanic Art* (Kungl. Vitt. Hist. och Antik. Akad., Stockholm, 1955); *Guldskatter från järnåldern* (State Historical Museum, Stockholm, 1959).

15 Bridget and Raymond Allchin, *The Birth of Indian Civilization* (Penguin Books, London, 1959), 284f.

16 Bo Gyllensvärd, *Chinese Gold and Silver in the Karl Kempe Collection* (Stockholm, 1953); also, *Bull. Mus. Far Eastern Ant.*, 29 (1957).

17 J. Edward Kidder, *Japan Before Buddhism* (Thames and Hudson, London, 1959), 168–70 and passim.

18 Richard E. W. Adams, *Prehistoric Mesoamerica* (Little Brown, Boston, 1977), 234, 270 & 309; Warwick Bray, 'Ancient American Metal-Smiths', *J.R.A.I.*, 1971, 25–43; and *op. cit.* (1978); H. Lechtman, *op. cit.*

19 Lucas, *op. cit.* (1962), 234f.

20 R. J. Forbes, *Metallurgy in Antiquity* (Brill, Leiden, 1950), 201f; Maryon and Plenderleith, *op. cit.*, 582–5.

21 Forbes, *op. cit.*, 172 & 190ff.

22 M. Rostovtzeff, *Iranians and Greeks in South Russia* (Oxford, 1922), pl. III.

23 Forbes, *op. cit.*, 197f.

24 Colin Renfrew, *The Emergence of Civilization* (London, 1972), 317–19.

25 E. Mackay, *The Indus Civilization* (London, 1935); B. and R. Allchin, *op. cit.*, 270, 285.

26 Gyllensvärd, *op. cit.*

27 J. A. Charles, 'The First Sheffield Plate', *Antiquity*, 42 (1968), 278–84. See also F. H. Stubbings, 'Appendix', *ibid.*, 14f.

28 H. Shetelig and H. Falk, *Scandinavian Archaeology* (Oxford, 1937), 194; J. Brøndsted, *Danmarks Oldtid* (Copenhagen, 1960), Band III, 170ff.

29 Warwick Bray, *op. cit.*, 36f.

30 *Impala Platinum Holdings Ltd Annual Report 1982*, 14f.

5 Precious stones

1 *Chambers's Encyclopaedia* (1950 edn), vol. 6, 199b; H. B. Walters, *Catalogue of the Engraved Gems and Cameos, Greek, Etruscan and Roman* (British Museum, 1926); John Boardman, *Greek Gems and Finger-Rings* (Thames and Hudson, London, 1979).

2 H. Frankfort, *Cylinder Seals* (Oxford, 1953).

3 K. E. Kluge, *Handbuch der Edelsteinkunde* (1860); Max Bauer, *Precious Stones* (1896) (Dover Publications, 1968).

4 G. Herrmann, 'Lapis lazuli: The Early Phases of the Trade', *Iran*, 30 (1968), 21–57.

5 G. C. Lamberg-Karlovsky and M. Tosi, 'Shahr-i Sokhta and Tepe Yahya: Tracks on the Earliest History of the Iranian Plateau', *East and West*, 23 (1973), 21–53.

6 Wheeler, *The Indus Civilization*, 3rd edn (Cambridge University Press, 1968), 79f.

7 G. E. Mylonas, *Ancient Mycenae. The Capital City of Agamemnon* (London, 1957), 160f.

8 Sir Arthur Evans, *The Palace of Minos* (London, Macmillan, 1935), vol. IV, figs. 349–50.

9 *Bull. de Coresp. Hellénique*, 88 (1964), 776.

10 Edith Porada, 'Further Notes on the Cylinders from Thebes', *Am. J. Arch.*, 70, 194.

11 G. Karo, *Die Schachtgräber von Mykenai* (Munich, 1930).

12 Robert Webster, *Gems: Their Sources, Descriptions and Identification* (Butterworth, London, 1975), ch. 14.

13 T. W. Beale, 'Early Trade in Highland Iran: A View from a Source Area', *World Archaeology*, 5 (1973), 133–48.

14 C. G. Lamberg-Karlovsky, *Excavations at Tepe Yahya, Iran: Progress Report no. 1* (Peabody Museum, Harvard, 1970), pl. 38.

15 Rostovtzeff, *op. cit.*, 19.

16 Tobler, *op. cit.*, pl. LVIII & LIX.

17 Wheeler, *op. cit.* (1968), 80.

18 H. C. Beck, 'Etched Carnelian Beads', *Ant. J.* 13 (1933), 284–98.

19 K. R. Maxwell-Hyslop, *Western Asiatic Jewellery c. 3000–617 B.C.* (Methuen, London, 1971), pl. 45.

20 Hugh Tait, *Jewellery Through 700 Years* (British Museum, 1976), no. 381.

21 *ibid.*, no. 383.

22 Robert Webster, *Gems in Jewellery* (N.A.G. Press, London, 1975), 55–60.

23 W. F. P. McLintock, *Gemstones in the Geological Museum*, 4th edn (HMSO, London, 1983), passim.

24 Tait, *op. cit.*, no. 6.

25 R. A. Higgins, *Greek and Roman Jewellery* (Methuen, London, 1961), 159ff.

26 Maryon and Plenderleith, *op. cit.*, ch. 23.

27 Rupert Bruce-Mitford, *The Sutton Hoo Ship Burial* (British Museum, 1972), pl. 32–4 and D, F; *The Sutton Hoo Ship Burial*, vol. 2 (British Museum, 1978), passim.

28 Lucas, *op. cit.*, 389f.

29 Higgins, *op. cit.*, 181f.

30 Joan Evans, *A History of Jewellery, 1100–1870* (London, 1953), 51.

31 S. Tolansky, *The History and Use of Diamond* (Methuen, London, 1962); E. Bruton, *Diamonds*, 2nd edn (N.A.G. Press, London, 1979).

32 Higgins, *op. cit.*, 181.

33 Hansford, *op. cit.*, 108–10.

34 Lucas, *op. cit.*, 193–4.

35 M. J. Hughes, 'A Technical Study of Opaque Red Glass of the Iron Age in Britain', *Proc. Prehist. Soc.*, 1972, 98–107.
36 Tait, *op. cit.*, p. 86.
37 *ibid.*, p. 158.
38 *200 Meisterwerke* (Kunsthistorisches Museum, Vienna, 1931), pl. 98.
39 Patrick J. Kelleher, *The Holy Crown of Hungary*, pl. 98 (American Academy in Rome, 1951).
40 Sutherland, *op. cit.*, pl. 168.
41 G. F. Kunz and C. H. Stevenson, *The Book of the Pearl* (Macmillan, London, 1908); Bauer, *op. cit.*, Appendix, 585–600.
42 *ibid.*, 589.
43 Jackson, *op. cit.*, 74–6.
44 *ibid.*, 80f.
45 Lamberg-Karlovsky, *op. cit.*, pl. 38 E.
46 Jackson, *op. cit.*, 118–21.
47 de Groot, *op. cit.*, 277–9.

6 The symbolic roles of precious substances

1 *Annual Report of the de Beer's Company for 1981.*
2 E. Steingräber, *Antique Jewellery. Its History in Europe from 800 to 1900* (Thames and Hudson, London, 1957), 176.
3 G. F. Kunz, *The Curious Lore of Precious Stones* (Dover, N.Y., 1971; original edn, 1913).
4 Joan Evans, *Magical Jewels of the Middle Ages and the Renaissance particularly in England* (Oxford, 1922), 15f.
5 *ibid.*, 24.
6 *Hist. Nat.* XVII, cap. II.
7 Williamson, *op. cit.*, 203–7.
8 Evans, *op. cit.*, 1922, 193.
9 *ibid.*, 35f.
10 Sir Jerome Horsey, *Travels* (Hakluyt Society, London, 1954), 200.
11 *ibid.*, 178.
12 Jackson, *op. cit.*, 1917, 91.
13 Evans, *op. cit.* (1922), 185.
14 Warwick Bray, *op. cit.*, 41.
15 *ibid.*, 12.
16 J. Brøndsted, *op. cit.*, 76ff; Ole Klindt-Jensen, *Gundestrupkedelen* (Copenhagen, 1961); P. V. Glob, *The Bog People* (Faber, London, 1969).
17 Proskouriakoff, *op. cit.*, 1974; Richard E. W. Adams, *Prehistoric Mesoamerica* (Little Brown, Toronto, 1977), 237f.

18 Warwick Bray, *The Gold of El Dorado* (Royal Academy, London, 1978), 18–23.

19 Richard Ellis, *Foundation Deposits in Ancient America* (Yale University Press, 1972), ch. 6.

20 The dates of reigns are quoted from J. A. Brinkman's Appendix to A. Leo Oppenheim, *Ancient Mesopotamia. Portrait of a Dead Civilization* (Chicago University Press, 1972).

21 Strong, *op. cit.*, xxv f.

22 Maxwell-Hyslop, *op. cit.*, lxi f.

23 W. H. Prescott, *History of the Conquest of Peru* (Everyman's Library, Dent, London, 1908), 58f.

24 Exodus, 28, 17–21. See also Ronald E. Clements, *The Cambridge Bible Commentary on the New English Bible* (Cambridge University Press, 1972), 79.

25 Joan Evans, *op. cit.* (1922), 79.

26 II Samuel, 12, 30.

27 E. T. T. (Lord) Twining, *A History of the Crown Jewels of Europe* (Batsford, London, 1960); *European Regalia* (Batsford, London, 1967). Both fully illustrated and indexed.

28 Martin Holmes and Maj.-Gen. H. D. W. Sitwell, *The English Regalia. Their History, Custody and Display* (H.M.S.O., London, 1972), 28.

29 Horsey, *op. cit.*, 272f.

30 Sutherland, *op. cit.*, 142.

31 R. A. G. Carson, *Coins, Ancient, Mediaeval and Modern* (Hutchinson, 1962).

Index

Abydos, 54, 73
adze, Maori (with nephrite blade), **Plate B**
adze-blades, 9, 38, **39**, 46, 47, **48**
adzes: symbolic role, 46, **Plate B**
Aegean world, 15, 55, 60, 65, 68
Afghanistan, **66**, 67
Africa, *see* East Africa, Egypt, South Africa, West Africa
agate, 65, 70, 91
alabaster, 87
Alleberg collar, **58**
alloys, 3, 50, 59, 60, 62
amber, 5, 13, 26, 28–30, 31, 84, **Plate A**
America, *see* Belize, Brazil, Colombia, Costa Rica, Ecuador, Eskimos, Greenland, Meso-america, Mexico, Peru, USA
amethyst, 23, 54, 65, 67, 68, 83, 84, **Plate C**
amuletic wands, 13
amulets, 29, 39, 42, 57, 70, 83, 84, 85
An-Yang, 41, **42**
aquamarine, 73–4, 83
Arabs, 17, 19, 22, 26, 80
Archbishop Maximian's throne, Ravenna, 93, **94**
artists, 88, 100, 101, **Plate J**
Asia, *see* Afghanistan, Arabs, Burma, China, India, Iran, Iraq, Japan, Sri Lanka, USSR

Assur, 87
Assyria, Assyrians, 15, 17, 93
Aunjetitz, 29
Australasia, *see* Australia, New Guinea, New Zealand
Australia, 53, 54, 72, 76
axe-blades, 9, 47, **48**
axes, 34, 50
  symbolic status, 45
Aztecs, 33, 47, 69, 72, 85

badges of office, 78
Barberini Ivory, 17, 18
Barma Grande, Mentone, 23
beads, 2, 15, 23, 26, 27, 28, 29, 30, 31, 48, 69, 70
Belize, 48
belt fittings, 31, 43
Benin, 26, 27
beryl, 5, 73, 91
bi (perforated discs), 40, **41**
bird plumage, 13, 17, 47
Birka, 26, 32
books (sacred), binding of, 77, 88, **90**, 91
bracelets, **2**, 7, 8, 15, 19, 27, 31, 45, 54, 60, 68, 83
bracteates, 55, **57**
Brassempouy, 15, **16**
Brazil, 76, 77
Brazilian Star of the South diamond, 77
breast-plate, 80, 91
  see also pectorals

bronze, 19, 25, 26, 27, 29, 35, 44, 50, 51, 54, 60, 69, 77, 104
  currency, 102
Bronze Age, 55, 60, 65
bronze castings, 3
bronzes, 10, 17, 24
Brunn, 15
buffet plate, London Gold-smiths' Company, **105**
burial garments, **44**
burials, 7, 23, 24, 30, 37, 41, 49, 85, 86
  palaeolithic, **2**
  prestigious, 31
  royal, 44, 85
  ship, 73, 85
  Sunghir, USSR, **2**
Burma, 16, 37
buttons, 14, 31, 45
Byzantium, Byzantines, 12, 17, 77, 78

calligraphic and painting accessories, 17, 19, 45
Cambridge, Chancellor of the University, **Plate K**
cameos, 51, **71**, 96
carborundum, 38
carbuncle, 83, 91
carnelian, 23, 68, 69, 70, 84, 87, 91, 93, **Plates C and D**
Catherine II's mitre crown, 97, **98**
cave-dwellers, Palaeolithic, 9
Celtic craftsmen, 27, 77, 100
celts, flat, 45

chalcedony, 5, 10, 65, 67, 68, 70–2, 83, 84, 87; *see also under* agate, carnelian, onyx, sardonyx
chalice, Abbot Suger's, 91, **92**
champlevé technique, 51, 61
chariot embellishments, 19
charms (*magatama*), 37
chest ornaments, 46, 49, 78
  see also breast-plate *and* pectorals
Chichén Itzá, Yucatan, 49, 87
China, Chinese, 3, 9, 10, 12, 15, 17, 19, 21, 22, 24, 25, 27, 28, 30, 33, 34, 35–45, 56, 60, 67, 75, 80, 85
chloromelanite, 33, 34
chryselephantine work, 14, 19, **20**
chrysolite, 91
chrysoprase, 83
cire-perdue method: *see* lost wax casting
cloisonné work, 51, 67, 68, 69, 73, 74, 77, **Plate I**
clothing: decoration, 2, 23, 78, 80, **Plate J**
coffins, 51, 54, 70, **Plate H**
coinage, 25, 64, 102, 104
  bronze, 25, 102
  copper, 25, 101, 102
  gold, 101, 102
  silver, 101, 102
coins: Greek and Roman, **103**
collar (gold), Alleberg, Sweden, **58**
collars, 27, 55
Colombia, 51, 62, 87
colophony, 30
combs, 19, 91
complex societies: *see* stratified societies
conspicuous consumption, 4, 13, 46, 47, 53, 60
conspicuous waste, 4, 46, 47, 85, 86
copal, 30

copper, 3, 29, 50, 54, 60, 61, 63, 87
  currency, 25, 101, 102
  symbolic use, 50
coral, 13, 26–8, 78, 83, 85
  sources, 26–7
  status of, 27
coronation, 93, 96, 98, 100
coronets, 104
corundum, 5, 38, 75
Costa Rica, 47
cowrie shells, 23–6, 28, 32, 78
  association with fertility and sight, 23–4
  as currency, 24–6, 28, 101
  insertion into eye sockets, 23, **24**, **25**
  reproduction in metal, 23, 24, 26, **Plate C**
  sources, 23, 25–6
  status, 24
  symbolism, 23, 26
craftsmen, specialised, 9, 88, 100–1
Crete, 14, 19, **20**, 54
crosses, 77, 91
crown
  of Aachen reliquary bust of Charlemagne, **96**
  of Catherine II (Russian imperial), 81, 97, **98**
  of Charlemagne, 96
  consort's, 99
  German imperial, 77
  Holy, of Hungary (St Stephen's), 77, 81, **95**
  of Queen Alexandra, 63, 93
  of Queen Elizabeth the Queen Mother, 98
Crown jewels, British, **99**, **Frontispiece**
crowns, 75, 77, 81, 93, 96, 97, 98
  see also Imperial State Crown *and entries under* crown
crozier, bishop's, 91, **Plate G**
crucifixes, 28

Cullinan diamond, 98
cult figure, Ur, 68
cup
  (gold), Mycenaean, **55**
  (silver), Hoby, Denmark, **61**
  (silver), Tang dynasty, **62**
cups and tumblers, 43, 44, 91
cups (trophies), 104
currency, 24, 25, 26, 64, 101, 102
  see also coinage
Cuzco, 88
Czechoslovakia, 15, 72

dagger, Mycenaean, 51, **63**
daggers: *see* personal weapons
Dashur, 23
definition of roles in hierarchy of authority, 93
Denmark, 56, **61**, 86, 87, **Plate A**
Derby Trophy, **106**
desk-furniture, 17, 22, 45
diadems, 68, 73
diamonds, 5, 10–11, 12, 38, 63, 72, 74, 75, **76**, 77, 78, 82, 83, 84, 95, 96, 97, **98–9**, 101, **Frontispiece, Plate J**
  cutting, 10, 73, 76–7
  sources, 75–6
Dioscorides, 83
Dong-Son culture, 25
Donja Dolina cemetery, 26

Eanna Temple, Uruk, 70
ear ornaments, 45, 46, 48, 49, 78, 80
East Africa, 80
ebonite, 31
ecclesiastical accessories, 6, 91
  see also religious buildings and appurtenances
Ecuador, 62, 63
Egypt, Egyptians, 10, 12, 16, 17, 19, 23, 24, 54, 59,

Egypt, Egyptians (*cont.*)
67, 68, 69, 70, 73, 78,
Plates C, H and **I**
el Argar culture, 60
electrum, 23, 50, 59, 102,
103, **Plate C**
Elizabeth I, Queen, 97, **Plate J**
emeralds, 10, 12, 72, 73–4,
75, **79**, 83, 84, 88, 91,
101, **Frontispiece**
source, 74
emulation, 4, 5, 11, 82, 83,
85, 86, 87
enamel, 5, 12, 51, 77–8, 79,
90, 91, 95, 104
en cabochon, 73, 74, 77, **79**,
88, 89, 92, 95
Eskimos, 29, 38, 47
Europe, *see* Czechoslovakia,
Denmark, France,
Germany, Great Britain,
Greece, Hungary, Italy,
Sweden, USSR,
Yugoslavia
Evans, Sir Arthur, 68
excellence, 1, 3, 11, 83, 87,
105
exchange
media, 28
networks, 9

faience, 12
farming communities,
Neolithic, 9, 29, 34, 45
felspar, 68
figures, 19, 22, 28, 29, 51
figurines, 6, 15, **16**, 35, 47,
48, 60
see also 'Venus' figurines
Fischer, Heinrich, 34
fish-hooks, 80
Flomborn cemetery, 9
Football Association Cup,
106
foundation offerings, 48, 87
'Founder's Jewel', New
College, Oxford, 78, 79
Foy, Ste, 88, 89

France, **4**, 5, 15, 16, 23, 45,
88, **89**
Fu Hao, 15, 24, 35, 41, 42
furniture, 70
furs, 13

Gagarino, 15
galena, 59–60
gaming boards and pieces,
14, 19
garnet, 38, 67, 68, 72, 73, **74**,
75
Gavrinis, Brittany, 45
Germany, 9, 29, **96**
gifts, 14, 17, 19, 24, 28, 60,
61, 69, 83, 85, 86, 88
gilding, 3, 12
gilt, 93, 104
girdles, 23
glass, 12, 17
coloured, 26, 68, 69, 77,
Plates C, H and **I**
gold, 3, 4, **5**, 10, 14, 15, 17,
19, **20**, 23, 24, 28, 30,
31, 33, 46, 50–7, **55**, **57**,
58, 59, 60, 61, 62, 63,
68, 69, 70, 73, **74**, 75,
78, 81, 82, 83, 84, 85,
86, 87, 88, 90, 91, 93,
96, **103**, 104, **Plates D,
H, I** and **K**
currency, 101, 102
indicator of status, 44
sheet, 31, 50, 51, 88, 89,
91, 93
sources, 52, 53, 54
status, 50, 52
gold bullion, 12, 102
gold leaf, 91
gold rushes, 52, 53, 54, 76
goldsmithing, 19, 50, 51, 54,
55, 57, 60
goldsmiths, 50, 51, 54, 57,
59, 68, 73, 91, 100, 101
grave goods, 7, 8, 24, 35, 41,
42, 49, 67, 68, 75, 85
Great Britain, **32**, 56, 74, 79,
99, **105**, **106**, **Frontis-**

piece, **Plates G, J** and **K**
Great Sancy diamond, 100
Greece, 6, 14, 17, 19, **20**, 23,
54, **55**, 68
Greenland, 14
greenstone, 10, 35, 37, 46, 53
see also jade
Gregory the Great, 91
Grotte des Enfants, 23
Guatavita, Colombia, 87
Gundestrup cauldron, 56, **86**,
87

haematite, 65
hair-pins, 28, 45
halberd blades, 35, 39, 41
handles, 17, 19, 22
Han empire, 25, 35, 44
Harappan civilisation, 55,
60, 68, 69, 70
head-dresses, 27, 28, 49, 51,
67, 70
hei-tiki, 46
hierarchies of excellence,
104, 105
hierarchy
of precious substances, 62,
65, 67, 82, 91
social roles and status, *see*
stratified societies
Hissar, 69
historic jewels, use to
strengthen legitimacy, 96
hoe-blades, 39
horse furnishings, 19, 24, 51,
97
hsuan chi (barbed and
serrated discs), 40
Hungary, 77, 81, **95**
hunter-fishers, 29, 47
hunters, Palaeolithic and
Mesolithic, 8, 14, 15, 29

iconography
Christian, 19, 77, 81, 88,
91
religious, 68, 88
images, wooden, 24, 56

Imperial State Crown, 81, 99, **Frontispiece**
India, 10, 26, 36, 38, 72, 73, 75, 76, 85, 98
Indus basin: *see* Harappan civilisation
inlay, 10, 14, 19, 35, 46, 51, 69, 70, 73, 80, 93
insignia, 9, 10, 41, 45, 46, 51, 65, 77, 87, 101, 104
intaglios, 51, 96
Iran, 67, 69, 80
Iraq, 15, 17, 54, 59, 67–71, 85, 87, 93, **Plate D**
Italy, 93, **94**
ivory, 5, 8, 9, 13–20, 26, 30, 69, 88, **90**, 91, 93, **94**, **Plate G**
  attributes, 6, 14, 19
  fossil, 13, 15, 16
  origins, 13, 14, 16, 17, 18, 35
  symbolic purposes, 14, 15
  working of, 14, 19
ivory chessmen, Isle of Lewis, 14
ivory crozier, Gardar, Greenland, 14
ivory sculpture, 14, 15, 19, 20

jacinth, 83
jade, 5, 6, 9, 10, 17, 19, 21, 30, 33–49, **39**, **40–2**, **43**, **44**, **48**, 50, 52, 56, 60, 65, 75, 78, 85, 87; *see also* jadeite *and* nephrite
  association with rulers, 41
  as sacrifice, 42
  sources, 34, 35, **36**, 37, 45, 47
  status, 33, 35, 38, 81
  symbolic role, 9, 33, 35, 39–40, 41, 45, 46, 47
  therapeutic properties, 33, 43–4
  use in burial rites, 44
  working of, 38, 39, 46
jadeite, 33, 38, 47, 48, 49, 73, 85

sources, 30, 34–5, 37, 45
symbolic role, 45–6, 47
jades, 9, 24, 38, 41, 42, 87
  musical, 42
Japan, 37, 56, 63
jasper, 68, 91
Jericho: Natufian levels, 23, **24**
jet, 13, 30–2, **32**, 84
  sources, 30, 31
  status of, 31
jewellery, 7, 11, 12, 13, 27, 28, 31, 50, 51, 54, 60, 63, 65, 68, 69, 70, 72, 73, 77, 78, **79**, 80, 82, 83, 85, 88, 105, **Plates C** and **D**
jewels
  sureties for loans, 100, 101

Kakovatos, 30
Khafajah, 87
King Solomon's throne, 14, 93
Knossos, 68–9
Koh-i-Noor diamond, 63, 77, 98
Kostienki I, 15

lacquer bowls, 17
lapidaries, 83, 84
lapis lazuli, 5, 10, 15, 23, 30, 54, 65, 67–9, 70, 75, 87, 88, 91, 93, **Plates C** and **D**
  sources, 67, 68
  spread of, **66**
Laugerie-Basse, Dordogne, 23
La Venta, Tabasco, 47–8
lead, 60
legitimisation of power and authority, 9, 41, 87, 96, 99
Les Espélugues, Lourdes, 15
Lewis chessmen, 14
loot, 14, 17, 59

lost wax casting, 55, 56, 57, 61
Lo-yang, 62
Lungshan culture, 36
lunulae, 31, 51

magatama charms, 37
magico/medical applications of precious substances, 83–5
Maikop burial, 60, 69
malachite, 73, 83, 87
mammoth ivory, *see* ivory: fossil
Mané-er-Hroek, Brittany, 45
Maori, 34, 35, 37, 38, **39**, 46, 47, **Plates B** and **F**
Marbode, bishop of Rennes, 84
masks, 51, 63, 69, **Plate E**
Maya, 10, 47, 57
medals, 104
Mediterranean world, 17, 19, 72, 77, 93
Meiendorf, Germany, 29
Mentone, 9, 23
mere (short-handled club), 46
Mesoamerica, 47
Mesopotamia, 59, 65, 67, 68, 87
metallurgy, 3, 54, 56, 57, 59
  symbolic needs, 3, 56
Mexico, 47–9, **Plate E**
Meyer, A. B., 34
Mezine, 15
millefiori glass, 74
Minoan Crete, 14, 19, 54
Minoan snake goddess, Crete, 19, **20**
Mostagedda, 13
mother-of-pearl, 78, 80
Mycenaean culture, 28, 30, 51, **55**, **63**, 67, 68, 69

necklaces, 7, 8, 15, 23, 26, 27, 29, 31, **32**, 48, 49, 68, 70, 78, 84, **Plate C**
nephrite, 33–9, 45, 46

nephrite (*cont.*)
  as indicator of status, 46,
    Plates B and **F**
netsuke, 14
New Guinea, 23, **25**
New Zealand, 46, 47, **Plates B**
  and **F**
niello, 61
Nimrud, 15, 17
Nohmal, Belize, 48

objects of display, 82
objects of parade, 51
obsidian, 68
offerings: *see* foundation
  offerings *and* votive
  offerings
Olmec ceremonial centre, La
  Venta, 47, **48**
Olympic medals, 104, **106**
onyx, 65, 70–1, 91
opals, **5**, 72
orb, 97
orders of chivalry, 78, 104
Orloff diamond, 97

Palestine, 23, **24**
Peacock Throne, 93
pearls, **5**, 6, 10, 13, 17, 44,
    72, 75, 77, 78–81, **79**,
    85, 91, 92, 95, 97,
    Frontispiece, Plate J
  sources, 78, 80
  status, 78, 80–1
pearls, cultured, **5**, 12
pectorals, 51, 68, 69, **Plate I**
pendants, 7, 8, 42, 45, 47, 48
Pepin, King of the Franks, 93
personal ornaments, 2, 7,8,
    15, 22, 27, 29, 31, 32,
    47, 48, 60, 62, 75, 80
personal trinkets, 17, 56
personal weapons, 43, 61,
    65, 68, 69, 73, 100
Peru, 88
Plantagenet crown, 101
plate (gold and silver), 87,
  104
platinum, **5**, 10, 62–4, 93

precious substances
  aesthetic appeal, 3, 6, 9,
    33, 50, 65, 67, 72, 75,
    78, 82
  derivation from distant
    sources, 10, 14, 17, 21,
    23, 30, 35, 38, 39, 46,
    65, 67
  durability as component of
    value, 13, 14, 23, 24, 33,
    43, 51, 65, 77, 82
  economic use, 101
  hierarchy of, 62, 65, 82,
    91, 101, 102
  mineral origin, 9, 13
  organic origin, 8, 13
  rarity as component of
    value, 33, 34, 38, 39, 52,
    63, 82
  role, 7, 42, 43, 82, 86, 87
Predmost, 15
prestige networks, 70
private ostentation, 60
  see also conspicuous con-
    sumption
Pu-abi, Queen
  grave goods, 67, 70
purse (Sutton Hoo), 73, **74**

Ravenna, 93, **94**
Ravlunda bracteate, **57**
regalia, 6, 96, 97, 104
  royal, 27, 97, 98, 99
Regent diamond, 100
religious buildings and
  appurtenances:
  embellishment of, 88,
    91, 104
reliquaries, 77, 88
reliquary statue of Ste Foy, **89**
repoussé work, 51, 57, 61
rhinoceros horn, 13, 17, **21**,
  22, 26, 30
rings, 10, 36, 43, 45, 64, 70,
  82, 91
rituals, 41, 42, 77, 93
rock crystal, 65, 67, 68
Rome, Romans, 10, 17,
  52–3, 60, 73, 75, 78, 80,

93, 101, 102
rosaries, 27, 28
rubies, 10, 12, 75, **79**, 84
ruby, balas, 97, 99, **Frontispiece, Plate J**

sacrifices, 42, 47, 85, 87
St Peter, Rome, 93
Samsi-Adad I, 87
sapphires, 10, 72, 75, 84, 91,
  97, 99, **Frontispiece,
  Plate J**
sardonyx, 71, 91, 92
Sargon II of Assyria, throne
  of, 93
scarabs, 68, 70, 78
sceptres, 45, 97, 98, **99**
Schliemann, Heinrich, 69
sculpture, 15, **16**, 19, **20**
seals and seal-stones, 15, 17,
  19, 44, 65, 68, 69, 70
Sennacherib, 87
Sepik River skull, 23, **25**
serpentine, 35, 47
Shahr-i-Sokhta, 67
shellac, 30
shells, **5**, 8, 9, 13, 23
  use for eyes/teeth, 45, 69,
    Plate E
  see also cowrie shells
shield ornaments, 73
silk route, 17, 35, 36
silver, 3, 15, 17, 23, 24, 28,
  44, 50–1, 57–61, **61, 62**,
  63, 68, 82, 83, **86**, 87,
  91, 93, **103**, 104, **106**
  currency, 101, 102
  native, 59, 60
  sources, 59, 60
  working of, 61
silver gilt, 104
silvering, 3
silversmithing, 54, 60, 61, 86
silversmiths, 59, 61
simetite, 29
Sinú, 86
skulls (human)
  clay copies, 23, **25**
  with plastered faces, 23, **24**

slate, 87
smiths, 3, 59, 63
  see also goldsmiths *and* silversmiths
smoking gear, 28
Snettisham gold torc, 56, 86
snuff boxes, 17, 45
social hierarchy: *see* stratified societies
society, reflection of, 7, 11
South Africa, 52–4, 74–5, 98, **99**
Spondylus gaederopus shells, 8, 9
Sri Lanka, 16, 73, 75, 80
'standard', Ur, 68
Star of Africa diamonds, 63, 98, 99
states, 9, 52, 53
status, 26, 27, 30, 40, 41, 44, 45, 49, 51, 81, 82, 93, 105
steatite, 65
stratified societies, 9, 10, 11, 19, 27, 30, 42, 46, 47, 53, 68, 100
Suger, Abbot, 91, 92
Sumer, Sumerians, 10, 15, 17, 51, 54, 56, 60, 67, 68, 70, 73
Sumerian court jewellery, **Plate D**
Sunghir, USSR, **2**, 15
Sutton Hoo
  purse lid, **74**
  ship-burial, 73, 85
Sweden, 26, 32, **57**, **58**
Sweet trackway, Somerset Levels, 45
swords: *see* personal weapons
symbol, definition of, 1
symbolic roles of precious substances, 82–106
  cementing of family relationships, 83, 87
  concern of group for individual, 85
  defining of roles in hierarchy of authority, 93

designation of functions and grades, 104
embodying of sense of continuity, 104
influencing of unseen powers, 87
marking of jubilees and anniversaries, 83
symbolism
  in Palaeolithic art, 7
  relations between individuals, 6
  relations between rulers and retainers, 27, 41
  religious, 3, 88, 91, 100
  role of individuals in society, 6, 82
symbols, 1, 9
  of achievement, 4, 11, 78, 104, 105
  of emulation, 9, 82, 83
  of excellence, 3, 4, 11, 12, 50, 51, 82, 104, 105
  of fertility and plenty, 43
  of love and regard, 83
  of mourning, 31
  of office, 82, 87
  of power and authority, 3, 10, 41, 46, 61, 93, 97, 98, 102
  of prestige, 26, 61
  of regeneration, 81, 91
  of sovereignty and majesty, 11, 75, 93, 96, 100, 102
  of status, 3, 9, 27, 30, 40, 41, 44, 46, 49, 51, 82, 91, 93, 100, 105
  of strength and hardness, 10, 75
  of wealth and affluence, 10, 26, 75
synthetic gems, **5**, 12

temple of Apollo, Delos, 87
temple of the Sun, Cuzco, 88
Tepe Gawra, 54, 67, 69
Thebes, 23, 68
Theophrastus, 65
therapeutic properties of

precious substances, 28, 33, 43–4, 84–5
thrones, 14, 91, 93, **94**
  as symbols, 93
tiaras, papal, 98, 101
Tikal, Yucatan, 49
tiki of greenstone (Maori), **Plate F**
  see also hei-tiki
tin, 29, 50
toilet requisites, 22
topaz, 12, 83, 84, 91
torcs, 27, 55, 56, 86
tourmaline, 12
trade, 10, 17, 22, 26, 27, 28, 29, 31, 34, 68, 72, 74, 75
trade networks, 10, 17, 19, 26, 29, 30, 54, 68, 80
Treasury of Athena, 87
tribute, 14, 17, 22, 30, 59, 69
trophies, 104
tumbaga (gold-bronze), 50, 63
turquoise, 5, 10, 15, 17, 35, 36, 54, 67, 68, 69–70, 73, 91, **Plate E**
Tutankhamun
  breast ornaments, 69, **Plate I**
  coffin of gold, 54, **Plate H**
  grave goods, 14, 19, 85
  throne, 93

Upper Palaeolithic burials, **2**, 7, 23
Ur, 59, 69, 70, 71, 85
  Early Dynastic Royal Cemetery, 51, 54, 67–8, 69, 85
USA, 47, 102
USSR, **2**, 15, 60, 69, 81, 97, **98**

values, 1, 3, 105
  expression by symbols, 1, 82
value systems, 3, 102
Vapheio cup, 55
Veblen, Thorstein, 3, 4, 5, 85
'Venus' figurines, **16**

vessels, 13, 17, 28, 39, 44, 51, 59, 60, 69, 88
   drinking, 51, 54, 61
   ritual, 3, 24, 91
vestments, 88, 91
votive offerings, 19, 29, 47–8, 49, 60, 86, 87
vulcanite, 31

weapons, 69
   see also personal weapons
West Africa, 25–7
wine cups, 15, 21, 22, 35
wine flagons, 27

Yahya, 69, 80

Yrieix, St, 88
Yugoslavia, 26

zong (four-sided blocks), 39, **40**
zun (bronze wine containers), 22